ANALYSIS OF FLOW IN PIPE NETWORKS

ANALYSIS OF FLOW IN PIPE NETWORKS

Roland W. Jeppson

Professor and Head
Department of Civil and Environmental Engineering
College of Engineering
Utah State University
Logan, Utah

ANN ARBOR SCIENCE
P.O. BOX 1425 • ANN ARBOR, MICHIGAN 48106

P. O. Box 1425, Ann Arbor, Michigan 48106

Library of Congress Catalog Card No. 75-36280
ISBN 0-250-40119-3

Manufactured in the United States of America

Preface

This book is an outgrowth of an instructional manual that was developed for use in conducting semi annual three day workshops, at Utah State University, dealing with computer analyses of steady flows and pressures throughout large pipe networks. Financial support for the first such workshop, and development of the manual, was provided by the Quality of Rural Life Program at Utah State University which was funded by the Kellogg Foundation.

Analyses of flow rates and pressures in pipe networks are needed whenever significant changes in patterns or magnitudes of demands or supplies occur in municipal water or natural gas distribution systems. These changes occur whenever new residential subdivisions or industries attach to the existing system or new sources of supply are tapped. In the absence of such analyses to determine the performance of the existing system under the new demands accompanied by studies for providing the most economical alternatives for eliminating deficiencies, needlessly large investments are made for larger than necessary pipes, redundant lines or duplicate facilities. Partly because such errors are buried, the tax payers, or consumers seldom become aware of these mistakes. With presently (1975) available computer programs such analyses can be obtained for almost insignificant computer costs. Furthermore, the amount and nature of data needed in order for these computer programs to define the network are minimal.

In recent years a number of significant improvements have occurred in the methods and techniques used to analyze steady flow in large pipe networks. The intent of this book is to provide the professional hydraulic engineer as well as the college engineering student with material through which he can become familiar with the various digital computer methods of pipe network analyses. For self study by the practicing hydraulic or municipal engineers in updating themselves, numerous example problems are solved in detail throughout the book. Since learning is based on doing in addition to reading, problems are provided (with answers) at the end of each chapter. These problems allow the student to apply the principles covered in that chapter. The book is also intended to supplement standard fluid mechanics or hydraulic text books in college courses with greater emphasis on pipe network analysis than given in such texts. The book is designed for use in engineering courses dealing with municipal water supply and distribution. Finally, the book should find use in continuing education in short courses and workshops.

The first three chapters deal with fundamental principles of hydraulics, the foundation of network analyses, with direct application to pressure conduit flow. The coverage of these principles is only to the extent necessary to fully understand how the various systems of equations came about whose solutions give the flow rates and/or heads throughout large pipe networks. Following this background material, the three separate systems of equations in current use are described. In subsequent chapters methods for solving these systems of nonlinear equations are presented. Short versions of computer progrms, in the form of FORTRAN LISTINGS, are given in these later chapters to illustrate what computer logic provides solutions by various methods of analyses. Special emphasis is given to methods for handling pumps and pressure reducing valves (or pressure regulators), as an integral part of the network. Finally, in an appendix, a description of input data required by a more extensive computer program is provided to illustrate what is necessary in preparing data for computer analysis today.

I wish to thank my colleagues at Utah State University, and the participants of the pipe network analysis workshops in providing the stimulus for writing the book. Recognition is due Allen Davis for promoting his employer CH_2M-Hill to provide financial support in developing computer programs to efficiently handle large pipe networks. Special thanks are due the typist, Barbara South, who not only typed the original, but also set-up the material for printing by the publisher. From past research and technical reports I have learned to greatly appreciate her competence in producing technical materials from almost illegible handwriting.

Roland W. Jeppson

Contents

Chapter		Page
I	Fundamentals of Fluid Mechanics	1
	Introduction	1
	Fluid Properties	1
	Density	1
	Specific weight	2
	Viscosity	3
	Example Problems Dealing with Fluid Properties	4
	Conservation Laws	6
	Introduction	6
	Continuity	6
	Example Problems Applying Continuity	8
	Conservation of Energy (Bernoulli Equation)	11
	Example Problems Dealing with Conservation Laws	17
	Momentum Principles in Fluid Mechanics	22
II	Frictional Head Losses	25
	Introduction	25
	Darcy-Weisbach Equation	25
	Friction Factor for Laminar Flow	27
	Friction Factors for Turbulent Flows	29
	Example Problems Using the Darcy-Weisbach Equation	32
	Computer Use with Darcy-Weisbach Equation (the Newton-Raphson Method)	37
	Empirical Equations	40
	Comparison of Empirical Equation with Darcy-Weisbach Equation	42
	Example Problem Using Empirical Equations	42
	Exponential Formula	44
	Example Problems in Defining Exponential Formula from the Darcy-Weisbach Equation	45

Chapter		Page

III Minor Losses 47

Introduction..47
Bends, Valves, and Other Fittings..47
Entrances, Exits, Contractions, and Enlargements..50

IV Incompressible Flow in Pipe Networks..53

Introduction..53
Reducing Complexity of Pipe Networks..54

Series pipes..54
Parallel pipes..55
Branching system..56
Minor losses..56

Example Problem in Finding Equivalent Pipes..58
Systems of Equations Describing Steady Flow in Pipe Networks..59

Flow rates as unknowns..59

Example Problem in Writing Flow Rate Equations..63
Heads at Junctions as unknowns..65

Corrective flow rates around loops of network considered unknowns..67

V Linear Theory Method..71

Introduction..71
Transforming Nonlinear Energy Equations Into Linear Equations..71
Example Problems Using Linear Theory for Solution..79
Including Pumps and Reservoirs into Linear Theory Method Method..80
Including Pressure Reduction Valves Into Linear Theory Method..85
Example Problems Which Include Pumps and Reservoirs..93

VI Newton-Raphson Method..115

Introduction..115

Head-equation..117

Chapter **Page**

Example Problems Based on the H-Equations 121
Corrective Flow Rate Equations 124
Including Pressure Reducing Valves in Analyses Based on the ΔQ Equations 129

PRV's acting as check valves 132

Example Problems in Solving ΔQ-Equations 134

VII Hardy Cross Method 145

Introduction 145
Mathematical Development 145

Appendix A—Units and Conversion Factors
Unit Definitions (SI) 151
Appendix B—Areas From Pipe Diameters 153
Appendix C—Description of Input Data Required by Programs Based on the Linear Theory Method in Solving the Q-Equations and the Newton-Raphson Method in Solving the ΔQ-Equations 155

General Specifications 156
Detailed Data for Each Pipe 160
Solutions to Variations of Problem 161

Index 163

Chapter I
Fundamentals of Fluid Mechanics

Introduction

This chapter provides the necessary background information concerning the properties of fluids and the laws which govern their motion. Readers with some understanding of fluid mechanics might begin their reading with Chapter II. Those readers with better training in fluid mechanics, equivalent to that acquired by civil, mechanical, agricultural, or aeronautical engineers, might begin their reading with Chapter IV.

The first sectin of this chapter will introduct those properties of fluids which are needed to understand later material in this manual. Along with these properties you will become acquainted with symbols which are used to denote fluid properties, as well as units and dimensions which are associated with these symbols. Later these symbols will appear in equations, and consequently as you read this chapter you should make the association between the symbol and what it represents. After covering basic fluid properties, with particular emphasis on liquids (or more precisely referred to as incompressible fluids; gases are compressible fluids), the fundamental conservation principles of mass and energy are discussed in the later portion of this first chapter with special reference to liquid (i.e., incompressible) flow in pipes and other closed conduits.

Fluid Properties

Density

The mass per unit volume is referred to as the density of the fluid and is denoted by the Greek letter ϱ (rho). The dimensions of density are mass per length cubed or M/L^3. The English system of units (abbreviated ES) uses the slug for the units of mass, and feet for the unit of length. The dimensions commonly used in connection with the ES system of units are force, F, length, L, and time, T. In the Systeme internationale, SI, the common dimensions are mass, M, length, L, and time, T. Thus in using LFT, dimensions for the slug can be obtained by relating mass to force through the gravitational acceleration, g, i.e. F(force) = M(mass) x g(acceleration of gravity). Thus since the units of acceleration are ft/sec², the units of density are,

$$\frac{\text{Slug}}{\text{ft}^3} \text{ or } \frac{\text{lb-sec}^2}{\text{ft}^4}$$

in the English system.

In the international system (Systeme International d'Unites) of units (abbreviated SI) which is an outgrowth of the metric system, mass is measured in the unit of the gram, gr (or kilogram, kg, which equals 1000 grams); force is measured in Newtons, N, and length in meters, m. A Newton is the force required to accelerate 1 kg at a rate of 1 m/sec², and equals 10^5 dynes which is the force needed to accelerate 1 gr at 1 cm/sec².

The units of density in the SI system are:

$$\frac{\text{kg}}{\text{m}^3} \text{ or } \frac{\text{N-sec}^2}{\text{m}^4}$$

The conversion of density from the ES system to the SI system or vice versa can be determined by substituting the equivalent of each dimension as given in Appendix A. Upon doing this

$$1 \text{ Slug/ft}^3 = 515.363 \text{ kg/m}^3$$

The density of water equals 1.94 slug/ft³ or 1000 kg/m³ (1 gr/cm³).

Specific weight

The specific weight (sometimes referred to as the unit weight) is the weight of fluid per unit volume, and is denoted by the Greek letter γ (gamma). The specific weight thus has dimensions of force per unit volume. Its units in the ES and SI systems respectively are:

$$\frac{\text{lb}}{\text{ft}^3} \text{ and } \frac{\text{N}}{\text{m}^3} \text{ or } \frac{\text{kg}}{\text{m}^2\text{-sec}^2}$$

The specific weight is related to the fluid density by the acceleration of gravity or

$$\gamma = g\rho \quad \text{.......................(1-1)}$$

Since g = 32.2 ft/sec² (9.81 m/sec²), the specific weight of water is

$$\gamma = 32.2\,(1.94) = 62.4 \text{ lb/ft}^3 \quad \text{(ES)}$$

or

$$\gamma = 9.81\,(1000) = 9810 \text{ N/m}^3 \text{ (SI)}$$

Viscosity

Another important fluid property is its viscosity (also referred to as dynamic or absolute viscosity). Viscosity is the fluid resistance to flow, which reveals itself as a shearing stress within a flowing fluid and between a flowing fluid and its container. The viscosity is given the symbol μ (Greek symbol mu) and is defined as the ratio of the shearing stress τ (Greek letter tau) to the rate of change in velocity, v, or mathematically dv/dy. This definition results in the following important equation for fluid shear.

$$\tau = \mu \frac{dv}{dy} \qquad (1\text{-}2)$$

in which dv/dy is the derivative of the velocity with respect to the distance y. The derivative dv/dy is called the velocity gradient. Equatin 1-2 is valid for viscous or laminar flow but not for turbulent flow when much of the apparent shear stress is due to exchange of momentum between adjacent layers of flow. Later, means for determining whether the flow is laminar or turbulent are given.

From Eq. 1-2 it can be determined that the dimensions of viscosity are force multiplied by time divided by length squared or FT/L^2. Its units in the ES and SI systems are respectively,

$$\frac{\text{lb-sec}}{\text{ft}^2} \text{ or } \frac{\text{Slug}}{\text{ft-sec}} \text{ and } \frac{\text{N}\cdot\text{s}}{\text{m}^2} \text{ or } \frac{\text{kg}}{\text{m}\cdot\text{s}}$$

Occasionally the viscosity is given in poises (1 poise = 1 dyne sec/cm^2). One poise equal 0.1 N-sec/m^2.

Because of its frequent occurence, the absolute viscosity divided by the fluid density is separately defined and called the kinematic viscosity, ν (Greek letter nu). The kinematic viscosity thus is

$$\nu = \mu/\rho \qquad (1\text{-}3)$$

The dimensions of kinematic viscosity are length squared per time. Common units for ν in the ES and SI system respectively are:

$$\frac{\text{ft}^2}{\text{sec}} \text{ and } \frac{\text{m}^2}{\text{sec}}$$

Another often used unit for kinematic viscosity in the metric system is the stoke (or centistoke = 0.01 stoke). The stoke equals 1 cm^2/sec.

The viscosity of many common fluids such as water depends upon temperature but not the shear stress τ or dv/dy. Such fluids are called Newtonian fluids to distinguish them from non-Newtonian fluids whose viscosity does depend upon du/dy. Table 1-1 gives values of the absolute and kinematic viscosities of water over a range of temperatures.

Example Problems Dealing with Fluid Properties

1. A container with a volume of 8.5 ft³ (0.241 m³) is filled with a liquid. The container and liquid weigh 650 lbs (2890 m) and the container weighs 55 lbs (244.6 N). Determine the density and specific weight of the liquid.

Solution:

ES SI

Wt. of liquid = 650-55 = 595 lb;= 2890 - 244.6=2645.4N

$$\gamma = \frac{Wt}{Vol} = \frac{595}{8.5} = 70\ \frac{lb}{ft^3};\quad \gamma = \frac{2645.4}{0.241} = 11{,}000\ \frac{N}{m^3}$$

$$\rho = \frac{\gamma}{g} = \frac{70}{32.2} = 2.17\ \frac{Slug}{ft^3};\ \rho = \frac{11{,}000}{9.81} = 1120\frac{kg}{m^3}$$

Table 1-1. Viscosities of Water

Temperature		Viscosity, μ		Kinematic Viscosity, ν	
°F	°C	lb-s/ft² x 10^5	N·s/m² x 10^3 [a]	ft²/sec x 10^5	m²/s x 10^6 [b]
32	0	3.746	1.793	1.931	1.794
40	4.4	3.229	1.546	1.664	1.546
50	10	2.735	1.309	1.410	1.310
60	15.6	2.359	1.129	1.217	1.131
68	20	2.093	1.002	1.084	1.007
70	21.1	2.050	0.982	1.059	0.984
80	26.7	1.799	0.861	0.930	0.864
90	32.2	1.595	0.764	0.826	0.767
100	37.8	1.424	0.682	0.739	0.687

[a] 1 N·s/m² = 10 poises.
[b] 1 m²/s = 1 x 10^4 Stokes.

2. **Oil with $\mu = 2. \times 10^{-4}$ lb-sec/ft² flowing in a 4-inch (10.16 cm) diameter pipe has a parabolic velocity profile given by: $v = 0.45[1 - (r/R)^2]$ fps ($v = 0.1372[1 - (r/R)^2]$ m/s), in which r is the radial distance from the pipe centerline and R is the radius of the pipe. What is the shear stress in the oil at r = 0", 1" (2.54 cm) and 2" (5.08 cm)?**

Solution:

ES

$$\tau = -\mu \frac{dv}{dr} = \mu\,[0.9\,r/R^2]\,;$$

at

$$r = 0,\ \tau = 2. \times 10^{-4}\,(0.9)\left(\frac{0}{0.02778}\right) = 0 \text{ psf};$$

$$r = 1'',\ \tau = 2. \times 10^{-4}\,(0.9)\left(\frac{0.0833}{0.02778}\right) = 0.000540 \text{ psf};$$

$$r = 2'',\ \tau = 2. \times 10^{-4}\,(0.9)\left(\frac{0.1667}{0.02778}\right) = 0.00108 \text{ psf};$$

SI

$$\tau = \mu\,[0.2744\,r/R^2]$$

$$\tau = 0.009576\ (0.2744)\,(0/0.002581) = 0 \text{ N/m}^2$$

$$\tau = 0.009576\ (0.2744)\ (0.0254/0.002581) = 0.02585 \text{ N/m}^2$$

$$\tau = 0.009576\ (0.2744)\ (0.0508/0.002581) = 0.0517 \text{ N/m}^2$$

3. **With what force would the flowing oil in problem 2 tend to pull 1000 ft (304.8 m) length of the 4" (10.16 cm) pipe? Relate this force to the pressure drop in the pipe.**

Solution:

$$\text{Force } F_\tau = A_s\tau = (2\pi R \times \text{length})\tau$$

ES

$$F_\tau = 2\pi(0.1667)(1000)(0.00108) = 1.13 \text{ lbs};$$

SI

$$F_\tau = 2\pi(0.0508)\,(304.8)(0.0517) = 5.03 \text{ N}$$

A control volume of the 1000 ft of fluid shows that the difference in pressure forces at the ends must equal the shear force, i.e.

$$F_T = (P_1 - P_2) A = \Delta P A, \Delta P = F_T/A$$

ES

$$\Delta P = 1.13/(0.02778\pi) = 12.95 \text{ psf;}$$

SI

$$\Delta P = 5.03/(0.002581\pi) = 620 \text{ N/m}^2$$

Conservation Laws

Introduction

Many analytical computations in engineering and the physical sciences are based on a relatively few fundamental principles and concepts. Most important among these are Newton's laws of motion, and conservation of mass, energy, and momentum. With limited exceptions mass is not created or destroyed, energy is only converted from one form to another, and momentum is only changed by force acting through time. In fluid mechanics the quantification of the conservation of mass is referred to as the continuity equation in its various forms. An expression of the conservation of energy frequently used in hydraulics is the Bernoulli equation. The momentum principle is extremely useful in determining external forces acting on moving fluids. The equations obtained from these three conservation principles are the most fundamental equations used for solving fluid mechanics problems. These principles and equations will be discussed in this section with special reference to solving problems dealing with fluid flow in pipes.

Continuity

In more general forms, mathematical equations which embody the principle that mass is conserved are differential or integral equations. This is the case because in general the quantities which describe the flow, such as point velocities, are functions of positions in space and time. For flows whose variables vary with a single space coordinate, and do not vary in time, an algebraic equation can be used for describing the fact that mass is conserved. Flows that do not change in time are called steady flow, and if only one space coordinate is used the flow is one-dimensional. This manual deals only with steady, one-dimensional flows.

In solid mechanics the conservation of mass principle is applied by simply noting that the mass of any body remains constant. When dealing with fluid flows it is generally more convenient to deal with the amount of fluid mass passing a given section of the flow rather than to keep track of the positions of all individual particles of fluid. Thus the continuity equation will deal with mass flux passing a section of the flow instead of mass alone. Mass flux is simply the flow of mass or mass per time, and has the units of slugs/sec in the ES system and kg/sec in the SI system. Denoting mass flux by G it can be related to ρ, the area through which flow occurs and the fluid velocity by,

$$G = \rho AV = \rho Q \qquad \text{(1-4)}$$

in which A is the cross-sectional area of the section and V is the average velocity of the flow through the section. The area A is normal to the direction of the velocity. The symbol Q in the later part of Eq. 1-4 is the volumetric flow rate with dimensions of L^3/T and is given by,

$$Q = AV \qquad \text{(1-5)}$$

For steady flow in a conduit the mass flux through two sections denoted by subscripts 1 and 2 of pipe some distance apart must be equal if the flow is steady. Therefore

$$G_1 = G_2 \quad \text{or} \quad \rho_1 A_1 V_1 = \rho_2 A_2 V_2 \qquad \text{(1-6)}$$

Equatin 1-6 is one form of the continuity equation. For fluids that are incompressible, i.e. whose densities are constant regardless of the pressure, the continuity equation reduces to

$$Q_1 = Q_2 \quad \text{or} \quad A_1 V_1 = A_2 V_2 \qquad \text{(1-7)}$$

Occasionally the weight flow rate W is wanted. It equals gG or

$$W = gG = g\rho AV = \gamma AV \qquad \text{(1-8)}$$

In dealing with junctions of two or more pipes the continuity principle states that the mass flow into the junction must equal the mass flow out of the junction. Mathematically this principle is,

$$\Sigma G_i = \Sigma \rho_i Q_i = 0 \qquad \text{(1-9)}$$

in which the subscript i takes on the values for the pipes which join at the junction and the summation, Σ, indicates the sum of these G's with proper regard for sign. Again for incompressible flows Eq. 1-9 reduces to,

$$\Sigma Q_i = 0 \quad \ldots\ldots\ldots\ldots\ldots\ldots\ldots (1\text{-}10)$$

Equatins 1-9 or 1-10 will play an important role later in analyzing networks of pipes. Junctions in such networks are commonly referred to as nodes.

If the velocity is not constant throughout the flow section, calculus can be used to determine the mass, weight, or volumetric flow rates past a section. For example, if the volumetric flow rate in a pipe is to be determined, and the velocity in the pipe is a known function of the radial distance from the center of the pipe, then

$$Q = \int_A \upsilon(r)\, dA = \int_0^R \upsilon(r)\,(2\pi r dr)$$

in which R is the radius of the inside of the pipe. The average velocity V can be determined from this Q by

$$V = Q/A$$

Example Problems Applying Continuity

1. A 6-inch (0.1524 m) pipe is connected to an 8-inch (0.2032 m) pipe. If the average velocity V = 35 fps (10.67 m/s) in the 6-inch pipe what is the average velocity in the 8-inch pipe? Determine the mass flow rate, and the volumetric flow rate.

Solution:

$$A_8V_8 = A_6V_6, \quad V_8 = \frac{A_6}{A_8} V_6 = \left(\frac{D_6}{D_8}\right)^2 V_6$$

ES

$$V_8 = \left(\frac{6}{8}\right)^2 35 = 19.69 \text{ fps}$$

$$G = 1.94\,(35) \left[\frac{\pi}{4} \left(\frac{6}{12}\right)^2\right] = 13.33 \text{ slug/sec}$$

$$W = gG = 32.2\,(13.33) = 429.2 \text{ lb/sec}$$

$$Q = \frac{G}{\rho} = \frac{13.33}{1.94} = 6.87 \text{ ft}^3\text{/sec (or cfs)}$$

SI

$$V_8 = \left(\frac{0.1524}{0.2032}\right)^2 10.67 = 6.00 \text{ m/s}$$

$$G = 1000\,(10.67)\left[\frac{\pi}{4}(0.1524)^2\right] = 194.6 \text{ kg/sec}$$

$$W = 9.81\,(194.6) = 1909 \text{ N/sec}$$

$$Q = \frac{194.6}{1000} = 0.1946 \text{ m}^3\text{/sec}$$

2. **Four pipes join at a junction of a water distribution system, as shown. Withdrawal from the 6-inch (0.1524 m) pipe is 0.95 cfs (0.0269 m³/s), and measurements of flow rates in the two pipes flowing into the junction are 1.2 cfs (0.0340 m³/s), and 0.87 cfs (0.0246 m³/sec) respectively. What is the flow rate in the other pipe containing flow away from the junction? What is the average velocity in each pipe?**

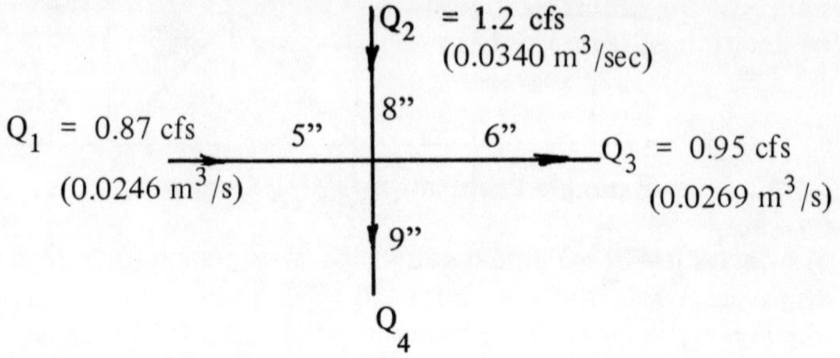

$$\Sigma Q_i = 0, \quad Q_1 + Q_2 - Q_3 - Q_4 = 0$$

ES

$$0.87 + 1.2 - 0.95 - Q_4 = 0$$

$$Q_4 = 1.12 \text{ cfs}$$

SI

$$0.0246 + 0.0340 - 0.0269 - Q_4 = 0$$

$$Q_4 = 0.0317 \text{ m}^3\text{/s}$$

$$V = \frac{Q}{A}$$

$$V_1 = \frac{0.87}{\frac{\pi}{144}(2.5)^2} = 6.38 \text{ fps}; \quad V_1 = \frac{0.0246}{\pi(0.0635)^2} = 1.94 \text{ m/s}$$

$$V_2 = \frac{1.2}{\frac{\pi}{144}(4)^2} = 3.44 \text{ fps}; \quad V_2 = \frac{0.0340}{\pi\,(0.1016)^2} = 1.05 \text{ m/s}$$

$$V_3 = \frac{0.95}{\frac{\pi}{144}(3)^2} = 4.84 \text{ fps}; \; V_3 = \frac{0.0269}{\pi(0.0762)^2} = 1.47 \text{ m/s}$$

$$V_4 = \frac{1.12}{\frac{\pi}{144}(4.5)^2} = 2.54 \text{ fps}; \; V_4 = \frac{0.0317}{\pi(0.1143)^2} = 0.772 \text{ m/s}$$

3. **A portion of a water distribution system is shown below in which the demands or consumptions of the system are given by the arrows. What total volumetric flow rate in cfs and gpm must the two reservoirs supply?**

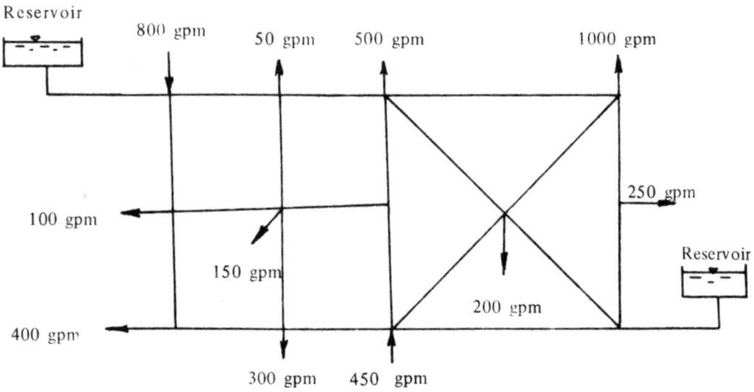

Solution:

Since the fluid is incompressible and steady, the total net inflow must be zero or,

$$\Sigma Q_i = 0; \; Q(2 \text{ reservoirs}) = 100 + 400 + 300 + 150$$
$$- 450 + 200 + 250 + 1000$$
$$+ 500 + 50 - 800 = 1700 \text{ gpm}$$

from Appendix A, 1 cfs = 449 gpm
Q = 1700/449 = 3.79 cfs

4. **Determine the volumetric flow rate and average velocity in a 2-inch (5.08 cm) diameter pipe if the velocity is given by $\upsilon = 0.1\,[1 - (r/R)^2]$ fps in which R is the radius of the pipe and r is the radial distance from the centerline.**

Solution:

$$Q = \int_0^{\frac{1}{12}} v\,(2\pi r dr) = 0.2\pi \int_0^{\frac{1}{12}} \left(r - \frac{r^3}{R^2}\right) dr$$

$$= 0.2\pi \left[\frac{r^2}{2} - \frac{r^4}{4\left(\frac{1}{12}\right)^2}\right]_0^{\frac{1}{12}}$$

ES	SI
$Q = 0.00109$ cfs	$Q = 0.0000309\ m^3/s$

$$V = \frac{Q}{A} = \frac{0.00109}{\pi(1/12)^2} = 0.05 \text{ fps}; \; V = \frac{0.0000309}{\pi(0.0254)^2} = 0.0152 \text{ m/s}$$

Note that the average velocity V equals 1/2 the velocity at the centerline ($v_c = 0.1$ fps) in which $r = 0$.

Conservation of Energy (Bernoulli Equation)

Flowing water contains three forms of energy which are of interest in hydraulics. Due to its motion it possesses kinetic energy, and it contains two forms of potential energy; one by virtue of its elevation and the other by virtue of its pressure.

Energy can be defined as the ability to do work. Work results from force moving through some distance, and therefore energy has the dimensions of force x length (FL). In a fluid flow the amount of energy passing a section will increase with time simply because more fluid will have passed that section. Consequently, it is more convenient to deal with energy per unit mass of fluid, or $E/M = FL/M$. In the ES, energy per unit mass has the units of lb-ft/slug, and in the SI the units are N•m/kg. The energy per unit mass due to the elevation of the fluid is gz, in which z is the vertical distance above some datum and g is the acceleration of gravity. The energy per unit mass resulting from the pressure is p/ϱ, in which p is the fluid pressure. The kinetic energy per unit mass of fluid is $V^2/2$. The sum of these three energies is,

$$E = gz + p/\rho + V^2/2$$

$$\text{(energy/unit mass)} = \begin{matrix}\text{due to}\\ \text{elev.}\end{matrix} + \begin{matrix}\text{due to}\\ \text{pressure}\end{matrix} + \text{kinetic} \qquad \text{(1-11)}$$

Each of these forms of energy will be examined in greater depth.

Potential energy due to elevation per unit mass. Consider a mass of fluid with its center of mass a distance z above an elevation datum as shown in Fig. 1-1.

With respect to the datum this mass possesses potential energy equivalent to its ability to do work in falling through the distance z. This work equals its weight (i.e. its mass times the acceleration of gravity) multiplied by the distance z or gMz. To get the energy per unit mass, gMz is divided by the mass M giving gz, or the first term on the right of Eq. 1-11.

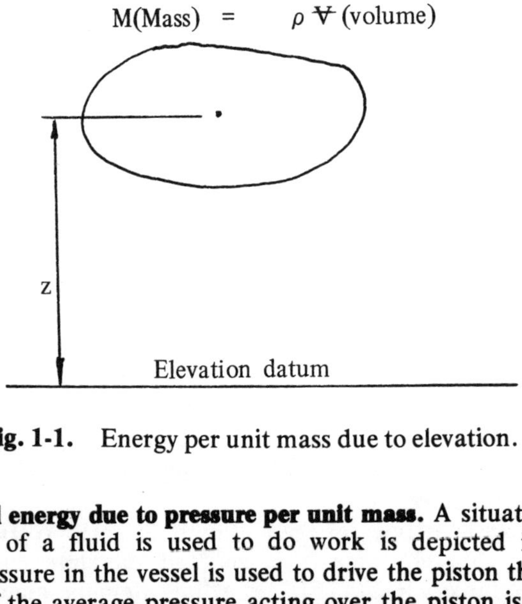

Fig. 1-1. Energy per unit mass due to elevation.

Potential energy due to pressure per unit mass. A situation in which the pressure of a fluid is used to do work is depicted in Fig. 1-2. The fluid pressure in the vessel is used to drive the piston through some distance ℓ. If the average pressure acting over the piston is p, the fluid pressure force on the piston is pA in which A is the area of the piston. Thus the energy obtained from the fluid in doing the work in moving the piston through the distnce, ℓ, equals $pA\ell$. The mass of fluid doing this work is that contained in the cylinder between the initial and final positions of the piston, a distance ℓ apart, or $M = \rho A\ell$. Consequently the energy per unit mass of the fluid due to pressure equals $pA\ell/\rho A\ell = p/\rho$ or the second term on the right of Eq. 1-11.

Kinetic energy per unit mass. The kinetic energy of any mass M equals 1/2 the mass times the velocity squared, or $1/2\, MV^2$. To obtain the kinetic energy per unit mass this quantity is divided by M resulting in $1/2\, V^2$ or the final term in Eq. 1-11.

Total energy per unit mass. The three simple sketches (Figs. 1 through 3) help in visualizing the three forms of energy per unit mass that water flowing in a pipe contains. In using these in a conservation of energy equation it is necessary to account for what happens to the energy level

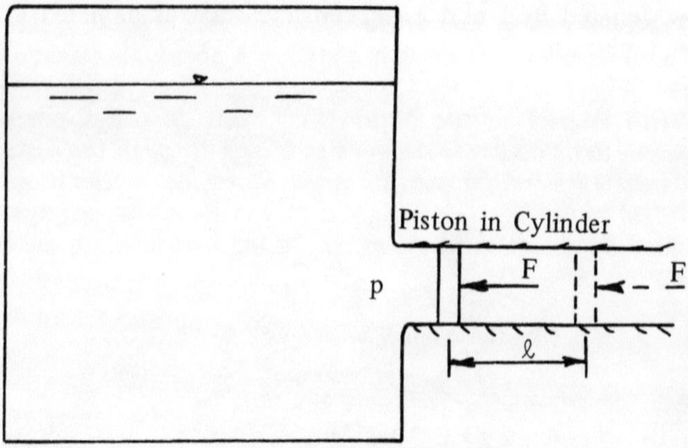

Fig 1-2. Energy per unit mass due to pressure.

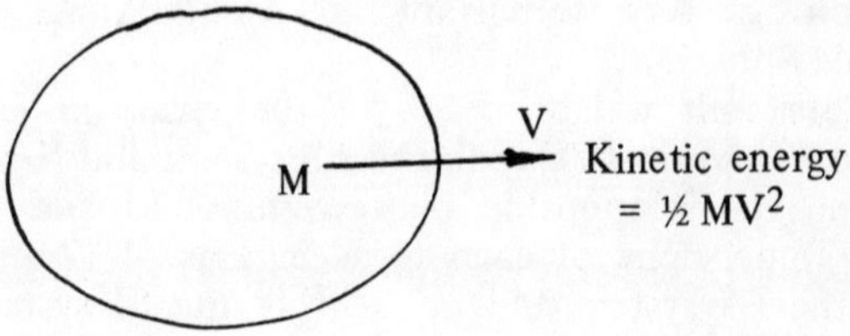

Fig. 1-3. Energy per unit mass due to velocity.

between two positions along the flow path or pipeline. Whenever fluid flow passes a fixed wall or boundary, fluid friction exists. This friction changes some of the useful flow energy into heat, or other forms of energy which are non-recoverable from the hydraulic viewpoint. In subsequent sections of this book methods for determining the magnitude of this lost energy per unit mass of flowing fluid will be discussed. For now it will simply be denoted by E_L.

A pump may exist in a pipeline which supplies energy to each unit mass of fluid passing through it, or a turbine may extract energy therefrom. These mechanical energies will be denoted by E_m, with the subscript standing for all forms of external mechanical energy. A pump produces a positive amount of E_m in the fluid and a turbine produces a negative E_m.

With these additional symbols for energy losses and all other forms of mechanical energy, the conservation of energy between two sections within a flow denoted by 1 and 2 respectively, such as depicted in the sketch below, is given by,

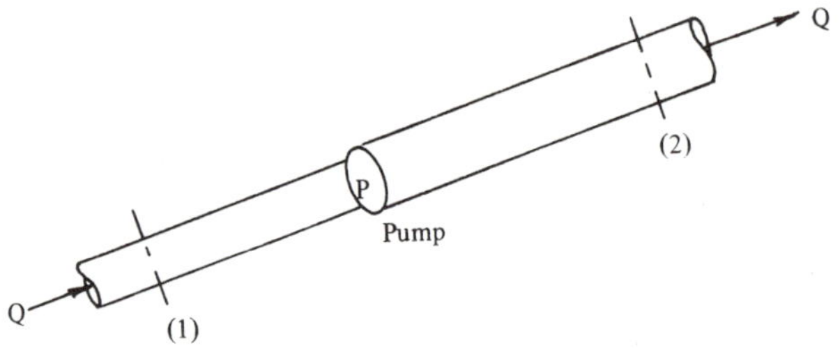

$$E_1 + E_m = E_2 + E_L \quad \ldots\ldots\ldots\ldots\ldots\ldots\ldots (1\text{-}12a)$$

Upon substituting from Eq. 1-11

$$gz_1 + p_1/\rho + V_1^2/2 + E_m = gz_2 + p_2/\rho + V_2^2/2 + E_L \ldots (1\text{-}12b)$$

This equation is very important in computations dealing with incompressible fluid flow.

Energy per unit weight. Each of the terms in Eqs. 1-12 have dimensions of FL/M (lb-ft/slug in the ES system and N•m/kg in the SI system). Using the relationship between mass M and force F, $F = ML/T^2$, the dimensions of each term in Eqs. 1-12 are also L^2/T^2 (ft^2/sec^2 in the ES system and m^2/sec^2 in the SI system). For many practical applications it is more convenient to express each term in the energy equation in dimensions of length L. This conversion of dimensions can be accomplished by dividing each term in Eq. 1-12b by g which has dimensions of L/T^2, giving

$$z_1 + \frac{p_1}{\gamma} + \frac{V_1^2}{2g} + h_m = z_2 + \frac{p_2}{\gamma} + \frac{V_2^2}{2g} + h_L \ldots\ldots\ldots\ldots\ldots (1\text{-}13)$$

in which $h_m = E_m/g$ and $h_L = E_L/g$

In hydraulic engineering practice Eq. 1-13 is used more widely than Eq. 1-12b, and is known as the Bernoulli equation, even though Eq. 1-12b might also be called the Bernoulli equation. These equations are very impotant in fluid flow, and therefore the meaning of each term should be fully understood. In Eqs. 1-12 each term represents energy per unit mass. Since Eq. 1-13 was obtained by dividing Eq. 1-12b by g and g multiplied by mass gives weight, each term in Eq. 1-13 represents energy per unit weight or FL/F = L.

Since the dimensions of each term in Eq. 1-13 is length, L (more correctly energy per pound), each such term is designated as a head; z is the elevation head; p/γ is the pressure head; and $V^2/2g$ is the velocity head. The sum of $z + p/\gamma$ is denoted as the piezometric or hydraulic head and the sum of $z + p/\gamma + V^2/2g$ is the total or stagnation head. Lines referred to as the hydraulic grade line (abbreviated HGL) and energy line (abbreviated EL) representing the piezometric and total heads, respectively, along a flow path are often drawn on sketches of the flow system. These lines not only convey considerable information about the flow but also aid considerably in understanding and analyzing a problem. The sketch of the system of pipes containing a turbine and a pump, shown below, illustrate the HGL and EL lines. In the reservoirs the velocity is zero, and on the water surface the pressure is zero (i.e. atmospheric), consequently the energy line is at the water surfaces in the reservoirs. The fact that the energy line at the upstream pipe entrance is slightly below the water surface in the reservoir is due to the entrance head loss, and the fact that the energy line enters the downstream reservoir above the surface means the velocity head above the water surface becomes the exit head loss. The amount that the EL drops between any two points in a pipe equals the head loss (or frictional loss) in that section of pipe. The amount of energy per pound of flowing fluid extracted by the turbine is represented by $h_m = -h_T$, and that added by the pump is $h_m = h_p$. The HGL is always the velocity head, $V^2/2g$, below the energy line. At the restricted section of pipe between the turbine and pump the velocity must increase to satisfy the continuity equation. With the resulting large increase in velocity head $V^2/2g$ at this section the pressure head, p/γ, must decrease, and for the situation depicted becomes negative, i.e. the HGL falls below the elevation of the pipe. At the entrance to the pump the pressure, which has become zero, is boosted through the pump to a relatively large value. The elevation head z is always taken as the distance between an arbitrarily selected datum and the fluid being considered, i.e., in this case the centerline of the pipe.

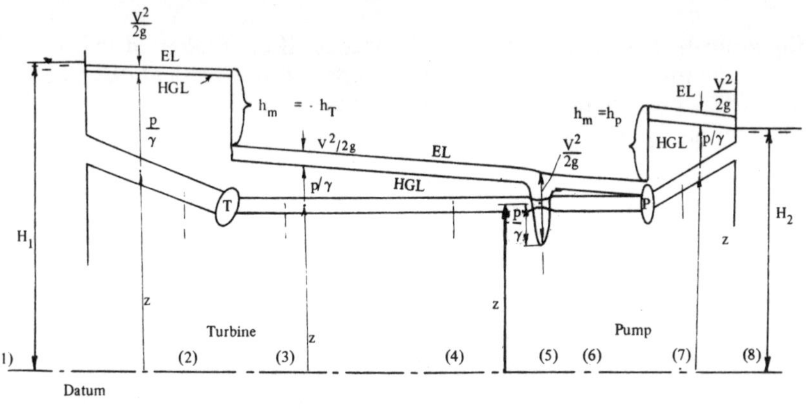

Fig. 1-4. Illustration of energy line (EL) and hydraulic grade line (HGL)

To illustrate the use of the Bernoulli equation, Eq. 1-13, it is written below between several of the sections identified by the numbers 1 through 8 on Fig. 1-4. You should practice writing it between other sections.

Between sections 1 and 2

$$H_1 = (h_L)_{entrance} + z_2 + p_2/\gamma + \frac{V_2^2}{2g} + h_{L_{12}}$$

Between sections 1 and 3

$$H_1 = (h_L)_{entrance} + z_3 + p_3/\gamma + \frac{V_3^2}{2g} + h_{L_{12}} + h_T$$

Between sections 3 and 5

$$z_3 + p_3/\gamma + \frac{V_3^2}{2g} = z_5 - \frac{|p_5|}{\gamma} + \frac{V_5^2}{2g} + h_{L_{35}}$$

Between sections 6 and 8

$$z_6 + p_6/\gamma + \frac{V_6^2}{2g} - h_{L_{68}} + h_P = H_2$$

In a resevoir or other body in which the velocity of the liquid is zero the Bernoulli equation reduces to

$$z_1 + p_1/\gamma = z_2 + p_2/\gamma \quad \text{(1-14)}$$

in which points 1 and 2 are any arbitrary points within the fluid. Solving Eq. 1-14 for

$$p_2 - p_1 = -\gamma(z_2 - z_1)$$

$$\text{or} \quad \Delta p = -\gamma \Delta z \quad \text{(1-15)}$$

Equation 1-15 is the equation of fluid statics. If z_1 is taken at the surface where the pressure is zero (i.e. atmospheric), and a vertical distance h is defined as positive downward from the water surface, Eq. 1-15 becomes the familiar equation of fluid statics,

$$p = \gamma h \quad \text{(1-16)}$$

which relates pressure to the depth below a liquid surface.

Equation 1-14 denotes that in a static liquid the sum of the elevation and pressure heads is constant. This constant is most easily determined at the liquid surface where the pressure is generally known, and often equal to atmospheric pressure which is taken as the zero pressure reference.

Conversion of energy per unit weight to power. Available information for pumps and turbines refers to power and efficiency and not the energy per unit weight of fluid passing through the device. Therefore equations are needed relating E_m to power P or horsepower HP. These equations are:

$$P = \frac{\gamma Q h_p}{\eta} = \frac{W h_p}{\eta} \quad \text{.................. (1-17a)}$$

and

$$HP = \frac{\gamma Q h_p}{550\eta} = \frac{W h_p}{550\eta} \quad ES \quad \text{......... (1-18a)}$$

$$\left(HP = \frac{\gamma Q h_p}{746\eta} = \frac{W h_p}{746\eta} \quad SI \right)$$

for pumps, and

$$P = (\gamma Q h_T)\eta = (W h_T)\eta \quad \text{................ (1-17b)}$$

$$HP = \frac{(\gamma Q h_T)\eta}{550} = \frac{(W h_T)\eta}{550} \quad ES \quad \text{....... (1-18b)}$$

$$\left(HP = \frac{(\gamma Q h_T)\eta}{746} = \frac{(W h_T)\eta}{746} \quad SI \right)$$

for turbines, in which η is the efficiency of the device and other terms are as previously defined.

Equations 1-17 and 1-18 can be derived by considering the dimesions of the quantities involved. Power is the rate of doing work or work per second. The energy exchange between the fluid and the device comprises the work and therefore energy lost or gained per second by the fluid should be equated to power after being modified by the efficiency of the device in carrying out this conversion. Since h_m (h_p for a pump or h_T for a turbine) is energy per unit weight, we can obtain power by multiplying h_m by the weight flow rate $W = \gamma Q$. The factors 550 and 746 in Eqs. 1-17 and 1-18, are the conversion of power to horsepower in the ES and SI systems respectively, since 1 hp = 550 ft-lb/sec (1 hp = 746 N•m/s).

Example Problems Dealing with Conservation Laws

1. The mass flow rate G of water through a pipe is 5.5 slugs/sec (80.267 kg/s). How much fluid energy flows through a section of this pipe in

---

[1] In the SI units Wh_p/η equal the number of watts of power. The watt (W) is the fundamental unit of power in SI. 1 W = 1 N·m/s.

10 minutes if at this section the pressure of the water is 25 psi (172.4 kPa) and its velocity equals 10 fps (3.048 m/s)?

Solution:

ES

$$\text{energy} = 5.5 \left(\frac{10^2}{2} + \frac{25(144)}{1.94}\right) 600;$$

$$= 6{,}290{,}000 \text{ ft-lb}$$

SI

$$\text{energy} = 80.267 \left(\frac{3.048^2}{2} + \frac{172.4 \times 10^3}{1000}\right) 600$$

$$= 8{,}527{,}000 \text{ N·m (joules)}$$

2. **The flow rate of a river is 1000 cfs (28.3 m³/s). A reservoir creates a depth of water 100 ft (30.48 m) above the intake to a power plant. Assume the hydraulic turbines are 80 percent efficient. How much potential power in horsepower and kilowatts can be generated at this site? (See Appendix A for conversion to watts.)**

$$HP = \eta\gamma QH/550$$

ES

$$\begin{aligned} HP &= 0.8\,(62.4)\,(1000)\,100/550; \\ &= 9{,}070 \text{ hp} \\ KW &= 9{,}070\,(0.746) = 6{,}770 \text{ kw} \end{aligned}$$

SI

$$\begin{aligned} HP &= 0.8\,(9800)\,(28.3)\,(30.48)/746 \\ &= 9{,}070 \text{ hp} \\ KW &= 9{,}070\,(0.746) = 6{,}770 \text{ kw} \end{aligned}$$

3. **Water for a fire hose is being pumped from a lake 5 ft (1.524 m) below the ground level and the fireman is holding the nozzle 3.5 ft (1.067 m) above the ground. If the pump increases the pressure by 30 psi (206.8 kPa), and the head loss in the fire hose is 4 ft (1.219 m), what velocity will the water have upon discharging from the nozzle? If the fireman held the nozzle vertically upward how high would the water rise (ignoring air drag)? If the nozzle is 1 inch (2.54 cm) in diameter what is the discharge and power required to drive the pump if it is 85 percent efficient?**

Solution:

Writing the Bernoulli equation, Eq. 1-13, between the lake and the nozzle, with the datum through the reservoir surface,

$$\overset{0}{\cancel{z_1}} + \overset{0}{\cancel{\frac{p_1}{\gamma}}} + \overset{0}{\cancel{\frac{V_1^2}{2g}}} + h_p = z_2 + \overset{0}{\cancel{\frac{p_2}{\gamma}}} + \frac{V_2^2}{2g} + \overset{0}{h_L}$$

in which $h_p = \Delta P/\gamma = 30 \times 144/62.4$

<u>ES</u>

$$\frac{30(144)}{62.4} = 8.5 + \frac{V_2^2}{64.4} + 4$$

$$V_2 = 60.4 \text{ fps}$$

<u>SI</u>

$$\frac{206.8 \times 10^3}{9810} = 2.591 + \frac{V^2}{19.62} + 1.219$$

$$V_2 = 18.4 \text{ m/s}$$

Height of rise above nozzle $= \dfrac{V_2^2}{2g}$

<u>ES</u>

$$\frac{(60.4)^2}{64.4} = 56.6 \text{ ft}$$

<u>SI</u>

$$\frac{(18.4)^2}{19.62} = 17.26 \text{ m}$$

Discharge $Q = VA$

<u>ES</u>

$$Q = 60.4 \left[\frac{\pi}{144} (0.5)^2 \right] = 0.329 \text{ cfs}$$

SI

$$Q = 18.4\ [\pi(0.0127)^2] = 0.00932 \text{ cms}$$

$$HP = \frac{\gamma Q h_p}{550\eta} = \frac{\gamma Q \Delta P/\gamma}{550\eta}$$

ES

$$HP = \frac{(0.329)\ 30 \times 144}{550\ (0.85)} = 3.04 \text{ hp}$$

SI

$$HP = \frac{(0.00932)\ (206.800)}{746(0.85)} = 3.04 \text{ hp}$$

4. **Water discharges at a rate of 1 cfs (0.0283 m³/s) from the pipe shown which leads from a reservoir with a water surface 40 ft (12.19 m) above the outlet. What is the head loss in the pipe and what is the pressure at the 1.8 inch (4.572 cm) section?**

Solution:

Writing Eq. 1-13 between the reservoir and the outlet,

$$H_1 = \cancel{z_2}^{\,0} + \cancel{p_2/\gamma}^{\,0} + V_2^2/2g + h_{L\,1\text{-}2} \quad \text{but } V_2 = Q/A_2$$

ES

$$V_2 = \frac{1}{\frac{\pi}{144}(1)^2} = 45.84 \text{ fps}$$

$$40 = \frac{(45.84)^2}{64.4} + h_L$$

$$h_L = 7.38 \text{ ft}$$

SI

$$V_2 = \frac{0.0283}{\pi(0.0254)^2} = 13.96 \text{ m/s}$$

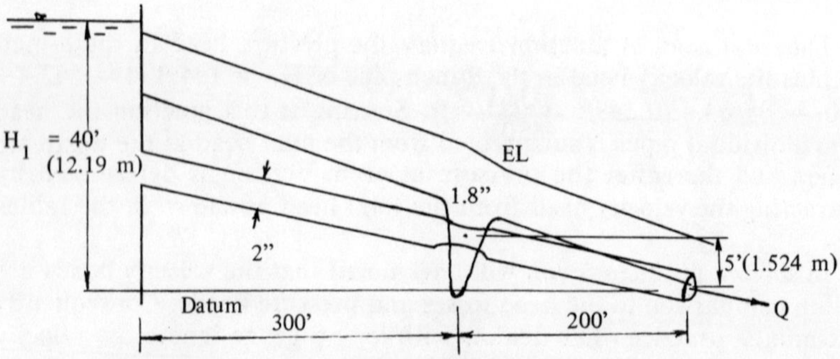

$$12.19 = \frac{(13.96)^2}{19.62} + h_L$$

$$h_L = 2.26 \text{ m}$$

Writing Eq. 1-13 between the reservoir and the section of 1.8 inch pipe:

$$H_1 = z_3 + p_3/\gamma + V_3^2/2g + h_{L_{1\text{-}3}}$$

The head loss $h_{L_{1-3}}$ equal approximately 3/5 of that determined above

ES

$$40 = 5 + \frac{p_3}{62.4} + \frac{(56.59)^2}{64.4} + 4.43$$

$$p_3 = -1195 \text{ psf}$$

SI

$$12.19 = 1.524 + \frac{p_3}{9810} + \frac{(17.23)^2}{19.62} + 1.356$$

$$p_3 = -57.1 \text{ kPa}$$

5. **A network of pipes has the external flows shown on the diagram. The table adjacent to the network gives the flows and head losses in the individual pipes. Determine the pressure at both ends of each pipe and the total head at each junction of the network if the pressure at junction [1] in the 8-inch pipe is 134.3 psi. All pipes are at the same elevation.**

Solution:

The total head at junction 1 equals the pressure head in the 8-inch pipe plus the velocity head in the 8-inch pipe or $H_1 = 134.3\ (144)/62.4 + (1660/449)^2/64.4\ (0.349)^2 = 311.7$ ft. Starting at this junction the head loss in individual pipes is substracted from the total head at the upstream junction and thereafter the pressure at each junction is determined by substracting the velocity head from the total head as shown in the tables below.

In solving problem 5 you will have noted that the velocity heads are small in comparison to the head losses and pressure heads. Consequently it is common practice when dealing with long pipes to ignore the velocity head and assume that the EL and HGL coincide.

Pipe	Q(gpm)	Head Loss (ft)
1-2	2,840	63.1
1-4	1,660	96.1
2-3	2,840	115.0
4-3	540	82.0
3-5	1,300	106.5
3-6	2,080	15.7
7-5	180	5.3
6-7	2,080	85.5

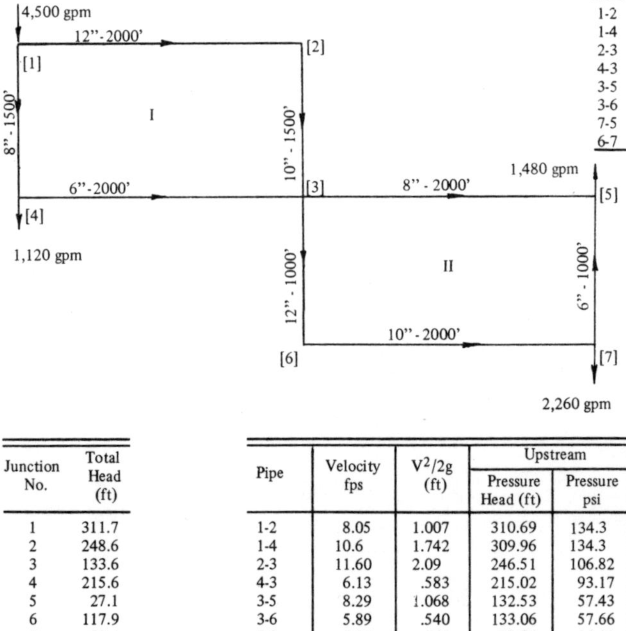

Junction No.	Total Head (ft)
1	311.7
2	248.6
3	133.6
4	215.6
5	27.1
6	117.9
7	32.4

Pipe	Velocity fps	$V^2/2g$ (ft)	Upstream		Downstream	
			Pressure Head (ft)	Pressure psi	Pressure Head (ft)	Pressure psi
1-2	8.05	1.007	310.69	134.3	247.59	107.3
1-4	10.6	1.742	309.96	134.3	213.86	92.67
2-3	11.60	2.09	246.51	106.82	131.51	56.99
4-3	6.13	.583	215.02	93.17	133.02	57.64
3-5	8.29	1.068	132.53	57.43	26.03	11.28
3-6	5.89	.540	133.06	57.66	117.36	50.86
7-5	2.04	.065	32.34	14.01	27.04	11.72
6-7	8.49	1.120	116.78	50.6	31.28	13.55

Momentum Principle in Fluid Mechanics

The third conservation principle, that of momentum, provides an additional powerful tool to solve many fluid flow problems, particularly those dealing with external forces acting on the fluid system, such as at elbows, junctions, and reducers or enlargers. A widely used equation resulting from the momentum principle is

$$\vec{F} = \rho Q (\vec{V}_2 - \vec{V}_1) \quad \ldots\ldots\ldots\ldots\ldots\ldots (1\text{-}19)$$

in which $\vec{F}$ is the resultant force (a vector with magnitude and direction) which acts on the fluid in a control volume being analyzed, $\vec{V}_2$ and $\vec{V}_1$ are the average velocities (also vectors) leaving and entering the control volume respectively.

The problems which are handled in this manual will not be solved by use of the momentum equation, Eq. 1-19, and consequently no additional explanation of its use is given here. Any text book dealing with fluid mechanics will provide additional treatment of this subject for the interested reader. The momentum equation will only be used in developing a small amount of theory given later.

Chapter II
Frictional Head Losses

Introduction

There are several equations which are often used to evaluate the friction head loss (i.e., conversion of energy per unit weight into a non-recoverable form of energy). The most fundamentally sound method for computing such head losses is by means of the Darcy-Weisbach equation. Discussion of the Darcy-Weisbach equation and related topics for computing frictional head losses will occupy the majority of this chapter. Use of the Hazen-Williams and Mannings equations will be described toward the end of the chapter.

Darcy-Weisbach Equation

The Darcy-Weisbach equation is,

$$h_f = \frac{\Delta p}{\gamma} = f \frac{L}{D} \frac{V^2}{2g} \quad \text{.................... (2-1)}$$

in which f is a dimensionless friction factor whose determination is described in the following pages, D is the pipe diameter, L is the length of pipe, V is the average velocity of flow, and g is the acceleration of gravity. For noncircular pressure conduits, D in Eq. 2-1 is replaced by 4 times the hydraulic radius, i.e. 4R. The hydraulic radius is defined as the cross-sectional area divided by the wetted perimeter or $R = A/P$.

The Darcy-Weisbach equation, Eq. 2-1, can be derived by dimensional analysis as is often done in books on the subject. This exercise will not be repeated herein since it serves only to demonstrate that Eq. 2-1 is rational. Rather a thumbnail sketch of extensive, systematic tests carried out by Nikuradse in about 1933 will be discussed to illustrate the functional dependency of the friction factor on two parameters, the Reynolds number Re and the relative roughness of the pipe wall e/D.

Nikuradse measured head loss, or pressure drops, caused by bonding uniform sand particles of various sizes, e, on the interior walls of different pipes. When his test results are plotted on log-log graph paper with the Reynolds number,

$$\mathrm{Re} = \frac{VD}{\nu} \quad \ldots\ldots\ldots\ldots\ldots\ldots\ldots\ldots (2\text{-}2)$$

plotted as the abscissa, and the friction factor f as the ordinate then data from different values of e/D defined the separate lines shown on Fig. 2-1.

The following conclusions can be drawn from Fig. 2-1: (1) When the Reynolds number Re is less than 2100, all data regardless of the magnitude of e/D fall on a single line defined by the equation f = 64/Re. Flow with Re <2100 is called *laminar flow* and Eq. 2-3 (below) is valid. (2) Data for different e/D values merge into the same line for smaller values of Reynolds number, but for Re> 2100 the flow is turbulent. For very smooth wetted pipe the data again fall on the same line. Flow conditions along this line have been classified as *turbulent smooth,* or *hydraulically smooth.* For these flows the majority of the flow is turbulent but the wall roughnesses are small enough to be embedded within the laminar sublayer. (3) For large Reynolds numbers and/or large values of e/D, the friction factor f becomes independent of Re, being only a function of e/D. For these flows the curves for different values of e/D on Fig. 2-1 are horizontal lines, and the flow is referred to as *"wholly rough"* or simply "rough." The previous three observations have lead to classifying flow as shown on Fig. 2-1 as *laminar, turbulent smooth, turbulent transitional,* or *turbulent rough.* The prefix "turbulent" is often omitted in referring to one of the three later types of pipe flow.

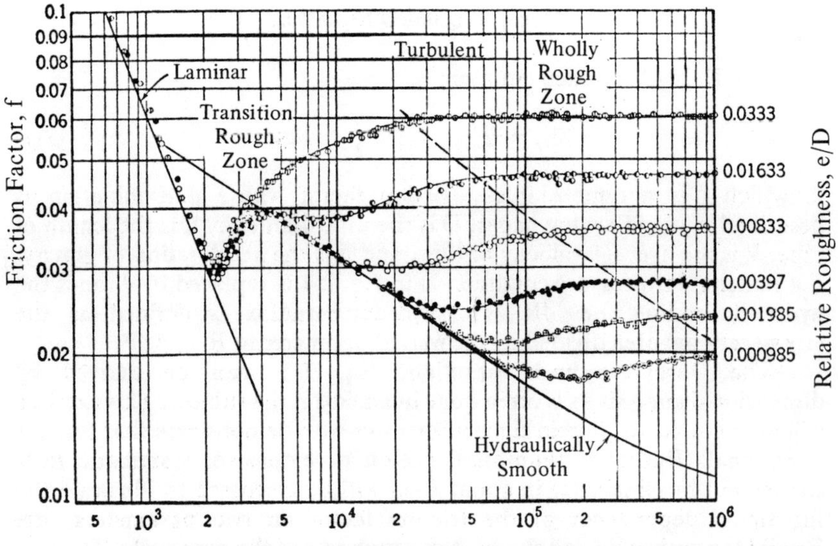

Fig. 2-1. Relationship of the friction factor to Reynolds number and the relative roughness of artificially roughened pipes.

Friction Factor for Laminar Flow

For laminar flows for which the well understood law of fluid shear, Eq. 1-2, applies, it is possible to provide a simple straight forward theoretical derivation of the Darcy-Weisbach equation, or more specifically derive the relationship

$$f = 64/Re \quad \text{.......................(2-3)}$$

For this derivation, note first that for a pipe, y in Eq. 1-2 can be replaced by R-r, so dy = -dr so

$$\tau = -\mu \frac{dv}{dr} \quad \text{.........................(2-4)}$$

This equation contains τ and v as unknown dependent variables as well as the independent variable r. Consequently, before it can be solved another equation is needed to provide a functional relationship between τ and v, or τ and r. This second equation comes from the momentum principle applied to a cylindrical element of fluid with length L and varying radius r within the pipe flow as shown in the sketch below. If this core of fluid is visualized as being isolated from the surrounding fluid it will have

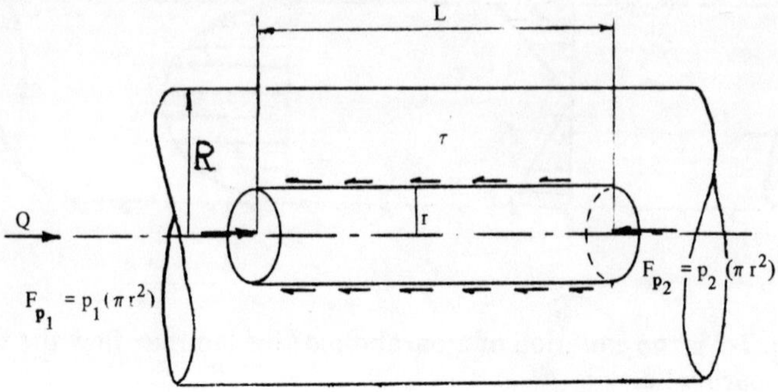

pressure forces p_1A and p_2A acting on its two ends and the force due to the shear stress τ between it and adjacent fluid acting on its cylindrical surface. Summing these forces for the left side of the momentum, Eq. 1-19 gives

$$(p_1 - p_2)\,\pi r^2 - \tau(2\pi r L) = \rho Q\,(V_2 - V_1) = 0$$

Since all forces and velocities are in the same direction for this application of the momentum, Eq. 1-19, only the magnitude of the vectors forces and velocities needs to be considered. Calling the pressure drop p_1-p_2, Δp and solving for τ yields

$$\tau = \frac{\Delta p}{2L} r \quad \ldots\ldots\ldots\ldots\ldots\ldots\ldots (2\text{-}5)$$

No assumption about whether the flow is laminar or turbulent was made in deriving Eq. 2-5; it is valid for all pipe flows. Since $\Delta p/2L$ does not change with r, Eq. 2-5 indicates that the shear stress τ varies linearly with r being zero at the pipe centerline where $r = 0$, and is a maximum at the pipe wall $r = R$. Upon substituting for τ from Eq. 2-5 into Eq. 2-4 and separating variables, the following first order ordinary differential equation is obtained

$$dv = -\frac{\Delta p}{2L\mu} r dr \quad \ldots\ldots\ldots\ldots\ldots (2\text{-}6)$$

which can readily be solved by integrating both sides. Upon integrating and evaluating the constant of integration from the boundary condition that $v = 0$ when $r = R$ gives,

$$v = \frac{\Delta p}{4L\mu}(R^2 - r^2) \quad \ldots\ldots\ldots\ldots\ldots (2\text{-}7)$$

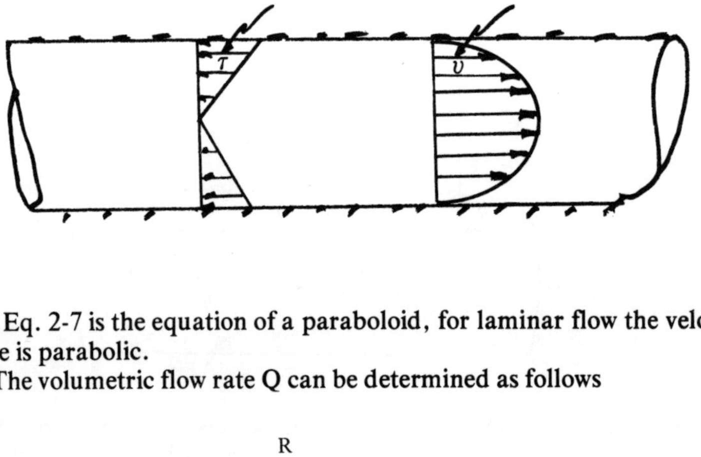

Since Eq. 2-7 is the equation of a paraboloid, for laminar flow the velocity profile is parabolic.

The volumetric flow rate Q can be determined as follows

$$Q = \int_0^R v\,(2\pi r dr) = \frac{\pi \Delta p}{8 L\mu} R^4 \quad \ldots\ldots\ldots\ldots (2\text{-}8)$$

and the average velocity

$$V = \frac{Q}{A} = \frac{\Delta p}{8L\mu} R^2 \quad \ldots\ldots\ldots\ldots\ldots\ldots (2\text{-}9)$$

Since the head loss equals the pressure drop divided by γ, h_f can be determined by solving Eq. 2-9 for Δp and dividing by γ, or

$$h_f = \frac{\Delta p}{\gamma} = \frac{8L\mu}{\gamma R^2} V \quad \ldots\ldots\ldots\ldots\ldots\ldots (2\text{-}10)$$

Equation 2-10 indicates that for laminar flow the frictional head loss h_f is proportional to the average velocity of flow since $8L\mu/(\gamma R^2)$ is constant for a given fluid in a given pipe.

Equating Eq. 2-10 to h_f in the Darcy-Weisbach equation, Eq. 2-1, and solving for f gives

$$f = \frac{64\mu/\rho}{VD} = 64/Re \quad \ldots\ldots\ldots\ldots\ldots (2\text{-}11)$$

which is identical to Eq. 2-3 obtained from Nikuradse's graph, Fig. 2-1, for laminar flow when $Re < 2100$. Thus theoretical analyses have verified the experimentally determined equation for the friction factor f for $Re < 2100$.

Friction Factors for Turbulent Flows

Equations relating f to Re and e/D for turbulent flow (i.e. flow with $Re > 2100$) cannot be obtained from similar elementary analyses, but are as summarized in Table 2-1, and discussed below.

The pipes used by Nikuradse were artifically roughened with uniform roughnesses and, therefore, cannot be applied directly to commercial pipes containing turbulent flows. Tests by others, notably Colebrook, demonstrated that flows in commercial pipes also become independent of Reynolds number, Re, at large Re and large wall roughnesses. Consequently, it is possible to compute the equivalent relative roughness e/D for commercial pipes from the experimental equation Nikuradse determined as valid for his wholly rough pipes. This equation for wholly rough pipes is (the last equation in Table 2-1):

$$\frac{1}{\sqrt{f}} = 1.14 - 2\log_{10}(e/D) \quad \ldots\ldots\ldots\ldots\ldots (2\text{-}13)$$

From these values of e/D, the equivalent sand grain size e for commercial pipes have been determined and are summarized in Table 2-2.

For "hydraulically smooth" pipes an analysis similar to that for laminar flow is possible but beyond the scope of this manual. This equation is (3rd equation in Table 2-1),

Table 2-1. Summary of Friction Factor Equations for Darcy-Weisbach Equation $h_f = f \frac{L}{D} \frac{V^2}{2g}$.

Type of Flow	Equation Giving f	Range of Application
Laminar	$f = 64/Re$ (Eq. 2-1)	$Re < 2100$
Hydraulically Smooth or Turbulent Smooth	$f = 0.316/Re^{0.25}$ (Eq. 2-15)	$4000 < Re < 10^5$
	$\frac{1}{\sqrt{f}} = 2 \log_{10}\left(Re \sqrt{f}\right) - 0.8$ (Eq. 2-14)	$Re > 4000$
Transition Between Hydraulically Smooth and Wholly Rough	$\frac{1}{\sqrt{f}} = 2 \log_{10}\left(\frac{e/D}{3.7} + \frac{2.52}{Re \sqrt{f}}\right) = 1.14 - 2 \log_{10}\left(\frac{e}{D} + \frac{9.35}{Re \sqrt{f}}\right)$ (Eq. 2-16)	$Re > 4000$
Hydraulically Rough or Turbulent Rough	$\frac{1}{\sqrt{f}} = 1.14 - 2 \log_{10} (e/D) = 1.14 + 2 \log_{10} (D/e)$ (Eq. 2-13)	$Re > 4000$

Table 2-2. Values of Equivalent Roughness e for Commercial pipes (New)

Material	e: Inches	e: cm
Riveted Steel	0.04 to 0.4	0.09 to 0.9
Concrete	0.01 to 0.1	0.02 to 0.2
Wood Stove	0.007 to 0.04	0.02 to 0.09
Cast Iron	0.0102	0.026
Galvanized Iron	0.006	0.015
Asphalted Cast Iron	0.0048	0.012
Commercial Steel or Wrought Iron	0.0018	0.046
PVC	0.000084	0.00021
Drawn Tubing	0.00006	0.00015

$$\frac{1}{\sqrt{f}} = 2 \log_{10}(\mathrm{Re}\sqrt{f}) - 0.8 \quad \text{..........(2-14)}$$

The friction factor f appears on both sides of the equation sign in Eq. 2-14, and consequently it cannot be solved explicitly for f with Re known, but must be solved by trial and error or some iterative scheme. An equation proposed by Blasius, which can be solved explicitly for f which apply to smooth pipes but only for flows with Re less than 10^5, is

$$f = \frac{0.316}{\mathrm{Re}^{0.25}} \quad \text{......................(2-15)}$$

Equation 2-14 applies to smooth pipe over the entire range of Re >4000, whereas Eq. 2-15 is an approximation to Eq. 2-14 limited to the range $4000 < \mathrm{Re} < 10^5$.

For the transition zone between smooth and wholly rough flow Colebrook and White give the following equation,

$$\frac{1}{\sqrt{f}} = 1.14 - 2 \log_{10}\left(\frac{e}{D} + \frac{9.35}{\mathrm{Re}\sqrt{f}}\right) \quad \text{........(2-16)}$$

Equation 2-16 gives nearly the same values for f as Eq. 2-14 for small values of e/D and values of f nearly equal to those of Eq. 2-13 for very large values of Re. Consequently, Eq. 2-16 may be used to compute f for all turbulent flows.

Particularly for hand computations it is convenient to summarize Eqs. 2-11 through 2-16 (the equations in Table 2-1) in a graph similar to

Fig. 2-1. This graph given as Fig. 2-2 has become known as the Moody diagram. Its use can eliminate the trial and error solution of Eqs. 2-14 and 2-16.

Example Problems Using the Darcy-Weisbach Equation

1. A flow rate of 150 gpm (0.00947 m^3/s) of oil with $\mu = 1.5 \times 10^{-3}$ lb-sec/ft^2 (0.0718 N•sec/m^2) and $\rho = 1.7$ slug/ft^3 (876 kg/m^3) occurs in a 4-inch (0.1016 m) pipe line. Determine the Reynolds number and classify the flow as laminar or turbulent.

Solution:

$$Re = \frac{VD}{\nu}, \text{ but } V = Q/A \text{ and } \nu = \mu/\rho$$

<u>ES</u>

$$Re = \frac{(150/449/(\pi/36))\,(1/3)}{1.5 \times 10^{-3}/1.7}$$

$$Re = 1{,}450$$

<u>SI</u>

$$Re = \frac{0.00947/(\pi \times 0.00258)\,(0.1016)}{0.0718/876}$$

$$Re = 1{,}450$$

Note Re is dimensionless and the result is independent of the unit system.

The flow is laminar since Re <2100.

2. What is the pressure drop and head loss per 1000 ft (304.8 m) for the flow in problem 1? What is the fluid shear stress on the wall of the pipe?

Solution:

The head loss or pressure drop can be computed from Eq. 2-1 with f = 64/Re or Eq. 2-10. Using Eq. 2-10, $\Delta p = (8L\mu/R^2)V$,

<u>ES</u>

$$\Delta p = \frac{8(1000)\,(1.5 \times 10^{-3})}{(2/12)^2}\,(3.828)$$

$$\Delta p = 1654 \text{ psf} = 11.5 \text{ psi}$$

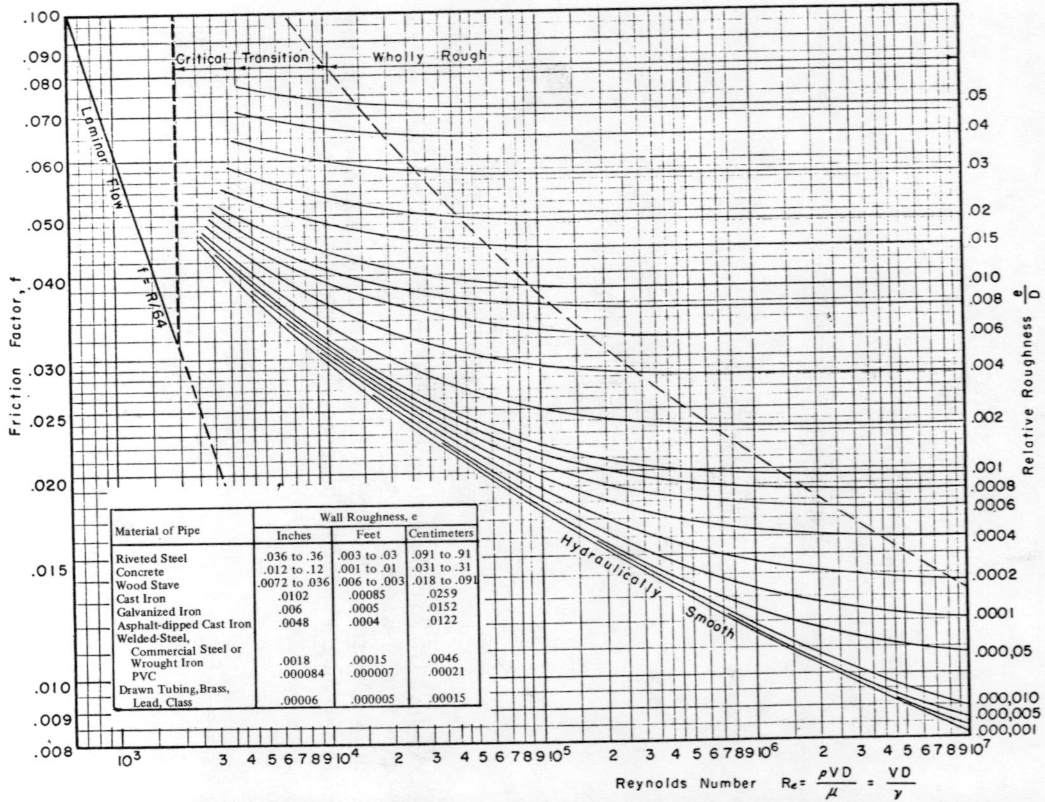

Material of Pipe	Wall Roughness, e		
	Inches	Feet	Centimeters
Riveted Steel	.036 to .36	.003 to .03	.091 to .91
Concrete	.012 to .12	.001 to .01	.031 to .31
Wood Stave	.0072 to .036	.006 to .003	.018 to .091
Cast Iron	.0102	.00085	.0259
Galvanized Iron	.006	.0005	.0152
Asphalt-dipped Cast Iron	.0048	.0004	.0122
Welded-Steel, Commercial Steel or Wrought Iron	.0018	.00015	.0046
PVC	.000084	.000007	.00021
Drawn Tubing, Brass, Lead, Class	.00006	.000005	.00015

Fig. 2-2. The Moody diagram giving f as a function of Re and e/D for commercial pipes.

$$h_L = \frac{\Delta p}{\gamma} = \frac{1654}{1.7(32.2)} = 30.2 \text{ ft}$$

SI

$$\Delta p = \frac{8(304.8)(0.0718)}{(0.0508)^2}(1.168)$$

$$\Delta p = 79{,}265 \text{ Pa}$$

$$h_L = \frac{79{,}265}{876(9.81)} = 9.22 \text{ m}$$

$$\tau_o = \frac{\Delta p}{2L} R$$

ES

$$\tau_o = \frac{1654}{2000}\left(\frac{2}{12}\right) = 0.138 \text{ psf}$$

SI

$$\tau_o = \frac{79{,}265}{609.6}(0.0508) = 6.61 \text{ Pa}$$

3. A 6-inch (15.25 cm) cast iron pipe 1500 ft (457.2 m) long is to carry a flow rate of water ($\nu = 1.217 \times 10^{-5}$ ft^2/sec) Q = 1.0 cfs (0.0283 m^3/s). Determine the friction factor (by equation and check with the Moody diagram) and the head loss in this pipe.

Solution:

ES	SI
e/D = 0.0102/6 = 0.0017	e/D = 0.259/15.25 = 0.0017

An estimate for f can be obtained from the equation for rough flow

$$\frac{1}{\sqrt{f}} = 1.14 - 2\log_{10}(e/D)$$

$$f = 0.0224$$

This value can be iteratively corrected by the equation for the transition zone

$$1/\sqrt{f} = 1.14 - 2\log_{10}(e/D + 9.35/Re\sqrt{f})$$

ES

$$Re = \frac{(1.0/0.1963)(0.5)}{1.217 \times 10^{-5}} = 2.09 \times 10^5$$

SI

$$Re = \frac{(0.0283/0.01824)\ (0.15250)}{1.131 \times 10^{-6}} = 2.09 \times 10^5$$

Iteration 1

ES

$$\frac{1}{\sqrt{f}} = 1.14 - 2 \log_{10} \left(0.0017 + \frac{9.35}{2.09 \times 10^5 \ \sqrt{0.0224}} \right)$$

$$f = 0.0234$$

SI

$$\frac{1}{\sqrt{f}} = 1.14 - 2 \log_{10} \left(0.0017 + \frac{9.35}{2.09 \times 10^5 \ \sqrt{0.0234}} \right)$$

$$f = 0.0234$$

Since the last iteration does not change f, f = 0.0234, which can be verified to two digits on the Moody diagram.

$$h_f = f \frac{L}{D} \frac{V^2}{2g} = 0.0234 \ \frac{1500}{0.5} \ \frac{(1./0.196)^2}{64.4} = 28.3 \text{ ft or } 8.63 \text{ m}$$

4. A 4-inch (10.16 cm) PVC pipe 6000 ft (1828.8 m) long is used to convey water at 68° F (20° C) between two reservoirs whose surface elevations differ by 150 ft (45.72 m). What is the flow rate Q?

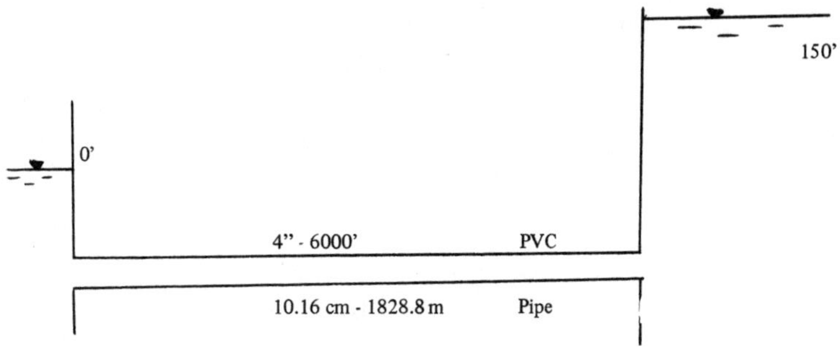

Solution:

The frictional hed loss equals 150' (45.72 m); therefore, write the Darcy-Weisbach equation with Q/A substituted for V

ES

$$150 = f \frac{(6000)\,Q^2}{(1/3)\,64.4\,(0.0873)^2}$$

$$fQ^2 = 0.00409$$

SI

$$45.72 = f \frac{(1828.8)\,Q^2}{(0.1016)\,(19.62)\,(0.008107)^2}$$

$$fQ^2 = 3.275 \times 10^{-6}$$

also $\quad Re = \dfrac{QD}{\nu A}$

ES

$$Re = \frac{0.333\,Q}{1.217 \times 10^{-5}\,(0.0873)} = 3.14 \times 10^5\,Q$$

SI

$$Re = \frac{0.1016}{1.131 \times 10^{-6}\,(0.008107)} = 1.108 \times 10^7\,Q$$

Solving by trial, Q is assumed, then Re is computed to enter the Moody diagram along the smooth pipe curve to obtain an f value, to examine if the equation is satisfied.

ES

Try $Q = 0.5,\ Re = 1.57 \times 10^5,\ f = 0.0165$

$fQ^2 = 0.0165\,(0.5)^2 = 0.00413$

$Q = 0.497,\ Re = 1.55 \times 10^5,\ f = 0.0165$

$fQ^2 = 0.0165\,(0.495)^2 = 0.00408$ O.K.

$Q = 0.497\ m^3/s$

SI

$Q = 0.014,\ Re = 1.57 \times 10^5,\ f = 0.0165$

$fQ^2 = 0.0165\,(0.014)^2 = 3.23 \times 10^{-6}$

Use $\quad Q = 0.0141\ m^3/s$

Computer Use with Darcy-Weisbach Equation (the Newton-Raphson Method)

Because the equations for determining f for smooth and transitional flow are implicit, requiring that problems solved by the Darcy-Weisbach equation be solved by trial, methods easily adapted to computer computations are described in this section. One very effective method for obtaining f in computer applications is to obtain an estimate of f from Eq 2-13 initially assuming rough flow, and then iteratively correcting this value of f by Eq. 2-16 as was done by hand in the previous problem (3). A computer algorithm implementing this approach is straight forward and therefore will not be discussed further. Rather a more general method, the Newton-Raphson method (often referred to as the "Newton method") will be discussed. This discussion will serve also as an introduction to Chapter VI of the manual dealing with methods for solving systems of equations describing pipe flow. At the present the Newton method will be used to solve a single implicit equation. Later it will be used to solve a system of nonlinear equations.

The Newton-Raphson method is an iterative scheme which starts with an estimate to the solution and repeatedly computes better estimates. It has "quadratic convergence." Other methods have linear convergence. Generally if quadratic convergence occurs, fewer iterations are needed to obtain the solution with a given precision than if linear convergence occurs. In addition to rapid convergence the Newton method is easily implemented in a computer algorithm. In using the Newton method the equation containing the unknown (which we will call x when describing the method in general), is expressed as a function which equals zero when the correct solution is substituted into the equation or $F(x) = 0$. For instance the friction factor Eq. 2-16 in the transition region would be written as,

$$F(f) = \frac{1}{\sqrt{f}} - 1.14 + 2 \log_{10} \left(\frac{e}{D} + \frac{9.35}{Re\sqrt{f}} \right) = 0 \quad \ldots\ldots(2\text{-}16)$$

as one form of $F(x) = 0$. The Newton method computes progressively better estimates of the unknown x by the formula,

$$x^{(m+1)} = x^{(m)} - F(x^{(m)})/dF^{(m)}/dx \quad \ldots\ldots\ldots(2\text{-}17)$$

in which the superscripts in parentheses are not exponents but denote number of iterations. To understand how Eq. 2-17 iteratively supplies better estimates for x consider a plotting of F(x) near a solution as illustrated below. When the initial guess $x^{(0)}$ is used in Eq. 2-17, $F(x) \neq 0$ since $x^{(0)}$ is not the correct solution. The derivative $dF(x^{(0)})/dx$ produces

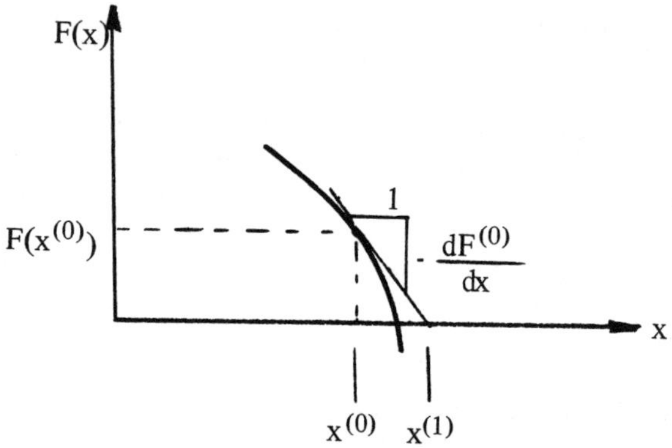

the slope of the curve shown on the sketch. If a line with this slope is projected to the x-axis a closer approximation $x^{(1)}$ to the solution is obtained. From similar triangles

$$\frac{x^{(1)} - x^{(0)}}{F(x^{(0)})} = -\frac{1}{\dfrac{dF^{(0)}}{dx}}$$

or solving for $x^{(1)}$

$$x^{(1)} = x^{(0)} - \frac{F(x^{(0)})}{\dfrac{dF^{(0)}}{dx}}$$

This is Eq. 2-17 with $m = 0$. If the procedure is repeated producing $x^{(1)}$, $x^{(2)}$, $x^{(3)}$ ... $x^{(m+1)}$ the function $F(x^{m+1})$ will approach zero, at which time the solution to the unknown x has been obtained.

The derivative dF/dx needed to solve Eq. 2-16 is

$$\frac{dF}{df} = -\frac{1}{2f\sqrt{f}} - \frac{9.35 \log_{10} \epsilon}{f\left(\dfrac{e}{D} + \dfrac{9.35}{Re\sqrt{f}}\right) Re\sqrt{f}} \quad \ldots\ldots (2\text{-}18)$$

In solving for f, Eq. 2-16 may be used whether the turbulent flow is smooth, transitional, or rough. However, since the Newton method does require an initial guess, and this can explicitly be supplied by the equation for trubulent rough flow, it is desirable to be able to distinguish rough flow from transitional flow without looking at a Moody diagram. A close approximation of the curve on the Moody diagram which separates these flows is

$$\frac{V\sqrt{f/8}}{\nu}\, e = 100 \qquad \ldots\ldots\ldots\ldots\ldots\ldots (2\text{-}19)$$

A computer program for solving for f for the Darcy-Weisbach equation should include the following features in the order listed: (1) Reading in the specifications such as D, e (or e/D) V (or Q or Re), ν, and L. (2) If Re is not a given specification then compute Re and test whether Re < 2100. If so f = 64/Re. Otherwise (3) compute an initial value for f from the rough equation, Eq. 2-13, (4) compute $(eV\sqrt{f/8})/\nu$ and if this quantity is greater than 100, then the f from step 3 is correct, otherwise (5) solve Eq. 2-16 by the Newton method. Listed below is a FORTRAN computer program which accomplishes this.

FORTRAN program for solving for the frictional head loss using the Darcy-Weisbach equation.

```
   10 READ(5,100,END = 99) D,Q,FL,VIS,E,G
C  D–pipe diameter, Q–flow rate, FL–length of pipe
C  VIS–kinematic viscosity of fluid x 10^5,
C  E–absolute roughness of pipe, G–acceleration of gravity
  100 FORMAT (8F10.5)
      A = .78539816*D*D
      V = Q/A
      RE = V*D/VIS
      IF (RE .GT. 2100) GO TO 3
      F = 64./RE
      GO TO 1
    3 EVIS = E/VIS
      ELOG = 9.35*ALOG10(2.71828183)
      ED = E/D
      F = 1./(1.14-2.*ALOG10(ED))**2
      PAR = V*SQRT(F/8.)*EVIS
      IF (PAR .GT. 100.) GO TO 1
      NCT = 0
    2 FS = SQRT(F)
      FZ = .5/(F*FS)
      ARG = ED + 9.35/(RE*FS)
      FF = 1./FS - 1.14 + 2.*ALOG10(ARG)
      DF = FZ + ELOG*FZ/(ARG*RE)
      DIF = FF/DF
      F = F + DIF
      NCT = NCT + 1
      IF (ABS(DIF) .GT. .00001 .AND. NCT .LT. 15) GO TO 2
    1 HL = F*FL*V*V*/(2.*G*D)
      WRITE (6,101) Q,D,FL,F,HL
  101 FORMAT ('Q=',F10.4,'D=',F10.4,'L=',F10.2,'F=',F10.5,'HEADLOSS='F10.4)
      GO TO 10
   99 STOP
      END
```

While the procedure described above for obtaining f by a computer is not difficult to implement, Wood[1] has provided an explicit approximation to the turbulent equation for f which simplifies the programming slightly at the expense of less precision. This approximation is,

$$f = a + b/Re^c \quad \ldots\ldots\ldots\ldots\ldots\ldots (2\text{-}20)$$

in which

$$a = 0.094\,(e/D)^{0.225} + 0.53\,(e/D)$$

$$b = 88\,(e/D)^{0.44}$$

$$c = 1.62\,(e/D)^{0.134}$$

Empirical Equations

While the Darcy-Weisbach equation is the most fundamentally sound method for determining head losses or pressure drops in closed conduit flow, empirical equations are often used. Perhaps the most widely used such equation is the Hazen-Williams equation, which is,

$$\text{ES} \qquad Q = 1.318\, C_{HW} A\, R^{0.63} S^{0.54}$$

$$\text{SI} \qquad Q = 0.849\, C_{HW} A\, R^{0.63} S^{0.54} \quad \ldots\ldots\ldots (2\text{-}21)$$

in which C_{HW} is the Hazen-Williams roughness coefficient, S is the slope of the energy line and equals h_f/L, R is the hydraulic radius defined as the cross-sectional area divided by the wetted perimeter, P, and for pipes equals D/4. Table 2-3 gives values for C_{HW} for some common materials used for pressure conduits.

If the head loss is desired with Q known the Hazen-Williams equation for a pipe can be written as

ES

$$h_f = \frac{4.73\,L}{C_{HW}^{1.852}\,D^{4.87}}\,Q^{1.852} \quad \text{with D and L in feet}$$

or

$$h_f = \frac{8.52 \times 10^5\,L}{C_{HW}^{1.852}\,D^{4.87}}\,Q^{1.852} \quad \text{with D in inches and L in feet}$$

$$\ldots\ldots\ldots\ldots\ldots\ldots (2\text{-}22)$$

---

[1]Wood, D. J., "An Explicit Friction Factor Relationship," Civil Engineering, ASCE, 36(12):60-61. December 1966.

Table 2-3. Values of the Hazen-Williams Coefficient, C_{HW} Manning's n for Common Pipe Materials

Type of Pipe	C_{HW}	n
PVC Pipe	150	0.008
Very Smooth Pipe	140	0.011
New Cast Iron or Welded Steel	130	0.014
Wood, Concrete	120	0.016
Clay, New Riveted Steel	110	0.017
Old Cast Iron, Brick	100	0.020
Badly Corroded Cast Iron or Steel	80	0.035

SI

$$h_f = \frac{10.7\ L}{C_{HW}^{1.852}\ D^{4.87}}\ Q^{1.852}$$

Another empirical equation is the Manning equation, which was developed for flow in open channels. The Manning equation is,

ES $$Q = \frac{1.49}{n}\ A\ R^{2/3}\ S^{1/2}$$

SI $$Q = \frac{1}{n}\ A\ R^{2/3}\ S^{1/2}$$

. . . . (2-23)

or for pipes with h_f on the left of the equal sign

ES

$$h_f = \frac{4.637\ n^2\ L}{D^{5.333}}\ Q^2 \quad \text{with D and L in feet}$$

or

$$h_f = \frac{2.64 \times 10^6\ n^2\ L}{D^{5.333}}\ Q^2 \quad \text{with D in inches and L in feet}$$

. . . . (2-24)

SI

$$h_f = \frac{10.29\ n^2\ L}{D^{5.333}}\ Q^2$$

Comparison of Empirical Equation with Darcy-Weisbach Equation

Since empirical equations such as Hazen-Williams equation (and the Manning equation to a lesser amount) are widely used for pipe flow computations, a comparison of their results with those of the Darcy-Weisbach equation are revealing. This comparison can best be done by manipulating the empirical equations into the form of the Darcy-Weisbach equation and then examining the relationships given by that equation in defining the friction factor f. Doing this with the Hazen-Williams equation using ES units produces.

$$f = \frac{6.05\,g}{C_{HW}^{1.852}\, D^{0.018}\, Re^{0.148}\, (\nu)^{0.148}} \quad \ldots\ldots\ (2\text{-}25a)$$

or using water at 60°F (15.6°C) ($\nu = 1.217 \times 10^{-5}$ ft²/sec) and $g = 32.2$ ft/sec².

$$f = \frac{1040.}{C_{HW}^{1.852}\, D^{0.018}\, Re^{0.148}} \quad \ldots\ldots\ (2\text{-}25b)$$

From Eqs. 2-25 it is evident that according to the Hazen-Williams equation f depends upon Re and to a lesser extent on the pipe diameter. For a given pipe size Eqs. 2-25 plot as a straight line on a Moody diagram. Fig. 2-3 contains six such lines superimposed on the Moody diagram for $C_{HW} = 100$, 130, and 150, respectively, and for D = 0.5 and 5.0 ft., respectively. The derivation of these straight lines from those lines for different e/D illustrates that the Hazen-Williams equation approximates the head loss only over a limited range of Reynolds numbers (or velocity) even if the appropriate value of C_{HW} is selected for a given Reynolds number (or velocity). The relationship of f from Mannings equation is,

$$f = \frac{5.75\, g\, n^2}{D^{1/3}} \quad \ldots\ldots\ (2\text{-}26)$$

Eq. 2-26 indicates that Manning's equation assumes the friction factor is independent of Re (i.e. in the wholly rough range), and depends upon n and the pipe diameter. Several lines obtained from Eq. 2-26 are also superimposed on Fig. 2-3

Example Problem Using Empirical Equations

1. Solve problem 3 of the previous problems for the head loss using the Hazen-Williams and Manning equations. (6" (15.25 cm) cast iron pipe, L = 1500' (457.2 m), Q = 1 cfs (0.0283 m³/s).)

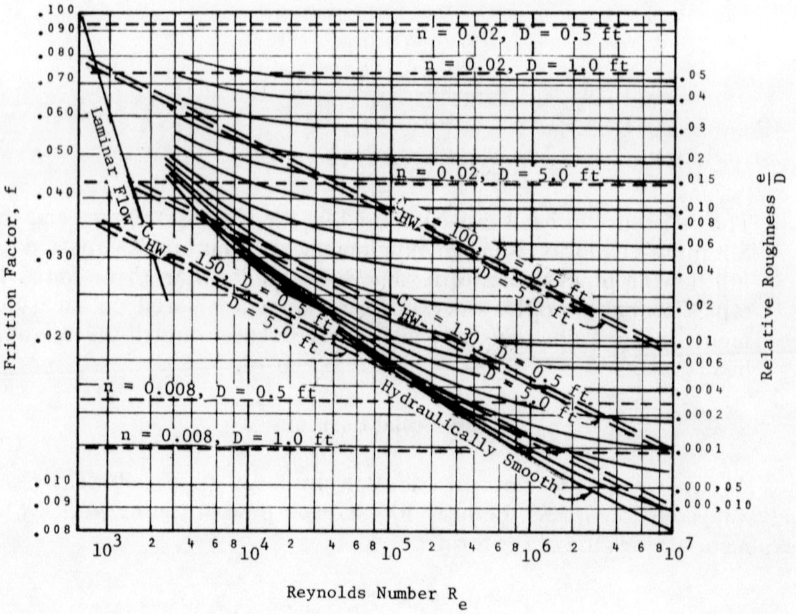

Fig. 2-3. Results from the Hazen-Williams and Mannings equations compared to the Darcy-Weisbach friction factor.

Hazen-Williams $C_{HW} = 130$

ES

$$h_f = \frac{8.52 \times 10^5 (1500)}{130^{1.852} \; 6^{4.87}} (1)^{1.852}$$

$$h_f = 25.3 \text{ ft}$$

SI

$$h_f = \frac{10.7(457.2)}{130^{1.852} \; (0.1525)^{4.87}} (0.0283)^{1.852}$$

$$h_f = 7.71 \text{ m}$$

Manning equation $n = 0.014$

ES

$$h_f = \frac{2.64 \times 10^6 \; (0.014)^2 \; (1500)}{6^{5.333}} (1)^2$$

$$h_f = 55.0 \text{ ft}$$

SI

$$h_f = \frac{10.29\,(0.014)^2\,(457.2)}{0.1525^{5.333}}\,(0.0283)^2$$

$$h_f = 16.8 \text{ m}$$

The difference in head loss computed by the Hazen-Williams and the Manning's equation is large. When using the Manning's equation for full flowing pipes values of n somewhat smaller than those found in tables probably apply, since tabular values are based on measurements in open channels in which surface waves contribute to some head loss.

Exponential Formula

In analyzing the flow distribution in large pipe networks it is advantageous to express the head loss in each pipe of the network by an exponential[1] formula of the form

$$h_f = K\,Q^n \qquad (2\text{-}25)$$

Values for K and n can be obtained directly from the previous equations given for the Hazen-Williams or Manning equations. To find K and n in Eq. 2-25 for the Darcy-Weisbach equation it should be noted that f can be approximated over a limited range by an equation of the form

$$f = \frac{a}{Q^b} \qquad (2\text{-}26)$$

Substituting Eq. 2-26 into the Darcy-Weisbach equation, Eq. 2-1, and grouping terms gives

$$n = 2 - b \qquad (2\text{-}27)$$

and

$$K = \frac{a\,L}{2\,gDA^2} \qquad (2\text{-}28)$$

Consequently determinations of n and K in the exponential formula, Eq. 2-25, requires finding values of a and b in Eq. 2-26 which cause that equation to approximate f for the range of flow rates to be encountered. If this range is too large n and K may be considered variables.

[1]The word "Power" formula describes the equation more appropriately. However, using "Power formula" in fluid mechanics would cause confusion with power expressed in watts or horsepower.

To illustrate how a and b might be determined, consider an 8-inch (0.2032 m) diameter welded steel pipe 500 ft (152.4 m) long carrying water at 60°F (15.6°C). The range of flow rates the pipe is to carry is 1.0 to 2.0 cfs (0.03 to 0.06 m^3/s), resulting in a velocity range of 3 to 6 fps (0.9 to 1.8 m/s), and values of Re from 1.6 x 10^5 to 3.2 x 10^5. With e/D = 0.0002, the friction factors from the Moody diagram for these two flow rates are f = 0.018 and 0.0165 respectively. Substitution of these two values for Q and f into Eq. 2-26 gives

$$0.018 = \frac{a}{1^b}$$

and

$$0.0165 = \frac{a}{2^b}$$

While a in this case can be obtained directly from the first of these two equations and b then from the second, in general they can be solved most readily by taking the logarithms of both sides or

$$\begin{aligned} -2 + 0.255 &= \log a - 0(b) \\ -2 + 0.217 &= \log a - 0.301\ b \\ \hline 0.038 &= 0.301\ b, \quad b = 0.126, \ a = 0.018 \end{aligned}$$

and therefore

$$n = 2 - b = 1.874$$

and

$$K = \frac{aL}{2gDA^2} = \frac{0.018\ (500)}{64.4\ (0.667)\ (0.349)^2} = 1.72$$

Example Problems in Defining Exponential Formula from the Darcy-Weisbach Equation

1. Cast iron pipes with e = 0.0102 inch (2.59 x 10^{-4} m) of the following lengths and sizes exist in a network. The range of velocities expected in these pipes is also given. Determine n and K for each of these pipes from the Darcy-Weisbach equation.

Pipe No.	Diameter		Length		V_1		V_2	
	Inch	m	ft	m	fps	m/s	fps	m/s
1	12	0.3048	700	213	0.425	0.130	0.851	0.259
2	10	0.254	1000	304.8	0.922	0.280	1.84	0.561
3	6	0.1524	600	182.9	5.10	1.55	10.21	3.11
4	8	0.2032	800	243.8	6.70	2.04	13.40	4.08

Solution:

Following the procedure given above K and n are:

Pipe No.	1	2	3	4
n	1.854	1.899	1.971	1.978
K (ES)	0.402	1.365	11.295	3.353
K (SI)	91.3	361	3850	1183

2. Repeat problem 1 using the Hazen-Williams equation with C_{HW} = 130.

Solution:

Pipe No.	1	2	3	4
n	1.852	1.852	1.852	1.852
K (ES)	1.84	1.90	1.97	1.98
K (SI)	97.3	315	2274	747

Chapter III
Minor Losses

Introduction

A number of appurtenances to pipelines such as inlets, bends, elbows, contractions, expansions, valves, meters, and pipe fittings commonly occur. These devices alter the flow pattern in the pipe creating additional turbulence which results in head loss in excess of the normal frictional losses in the pipe. These additional head losses are termed *minor losses.* If the pipelines are relatively long these losses are truly minor and can be neglected. In short pipelines they may represent the major losses in the system, or if the device causes a large loss, such as a partly closed valve, its presence has dominant influence on the flow rate. In practice the engineer must use professional judgment in deciding if and how many "minor losses" should be included in his analysis of fluid distribution systems. This chapter provides an introduction to determine magnitudes of minor losses.

Bends, Valves, and Other Fittings

The head loss (or conversion of flow energy into heat) due to a gradual bend in a pipeline is primarily caused by a secondary double spiral flow which is superimposed on the main flow. In moving around the bend the larger centrifugal force acting on the faster moving center core fluid forces it toward the outside of the bend as the slower moving fluid near the walls of the pipe moves toward the inside of the bend, as illustrated in the sketch below. This causes a double spiral or secondary motion which persists downstream as far as 50 to 100 pipe diameters until viscous forces eventually dissipate it. Thus the added head loss caused by the bend does not occur at the bend but over some length of pipe downstream therefrom.

Principles governing the flow of fluid as well as much experimental evidence indicates that the head loss due to added turbulence or secondary

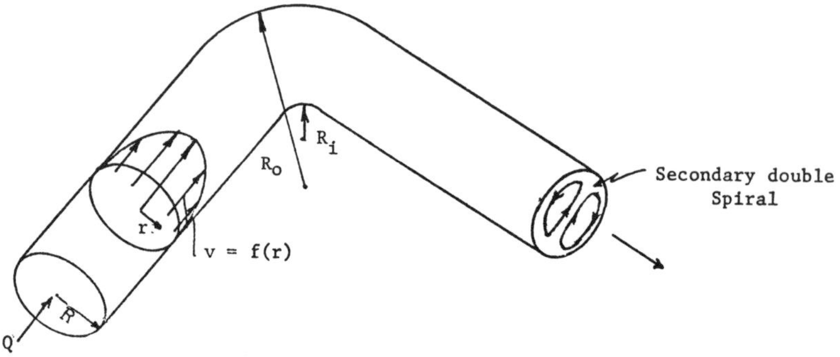

flows will be approximately proportional to the velocity squared or the flow rate squared. It is therefore common to express the majority of minor losses by an equation of the form,

$$h_L = K_L \frac{V^2}{2g} \qquad \text{(3-1a)}$$

or

$$h_L = C_L Q^2 \qquad \text{(3-1b)}$$

in which

$$C_L = K_L/(2gA^2)$$

Figure 3-1 gives the loss coefficient for gradual 90 degree bends as a function of the ratio of the radius of the bend divided by the pipe diameter and the relative roughness of the pipe wall. Figure 3-2 gives this coefficient related to the angle of bend.

For valves, sharp bends, and other fittings the minor head loss may be caused, to a much greater degree, by separation of the flow lines from the boundary at abrupt changes of wall geometry in the device rather than by secondary flows. Equation 3-1 also approximates such additional head losses provided an appropriate coefficient K_L is selected. Table 3-1 provides some nominal values of K_L for various common appurtenances to pipelines.

To illustrate how the magnitude of minor losses effects flow rates consider a 4-inch (0.1016 m) diameter cast iron pipeline which is 500 ft (152.4 m) long but contains a fully open globe valve. The flow rate which can be anticipated through this pipe is sought when the total head loss is known. Let us assume the available head loss and the minor loss due to the globe valve equals 30 feet, or

$$f \frac{L}{D} \frac{V^2}{2g} + K_L \frac{V^2}{2g} = \left(f \frac{L}{D} + K_L\right) \frac{V^2}{2g} = 30$$

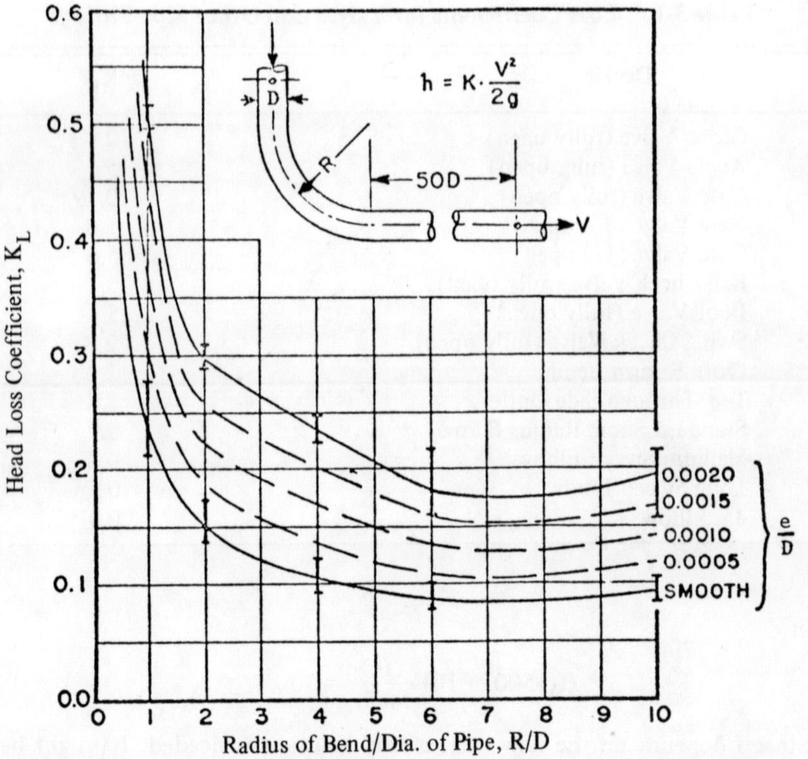

Fig. 3-1. Head loss coefficients for 90° bends of constant diameter.

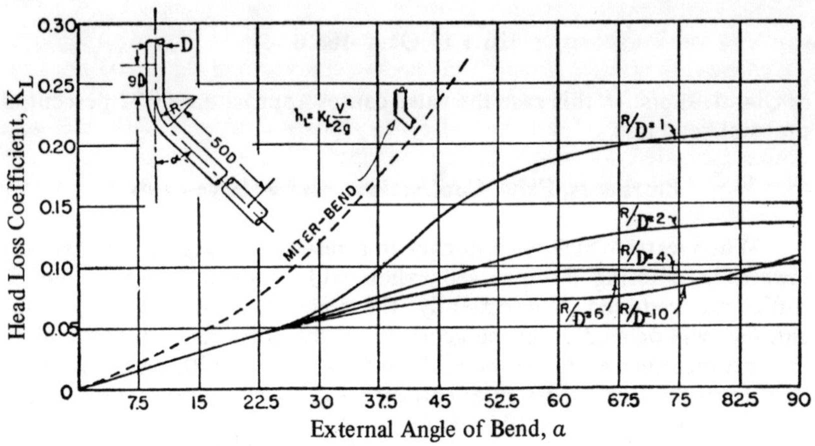

Fig. 3-2. Head loss coefficients related to degree of bend at Re $= 2.3 \times 10^5$.

Table 3-1. Loss Coefficients for Valves and Other Pipe Fittings

Device	K_L
Globe Valve (fully open)	10
Angle Valve (fully open)	5
Gate Valve (fully open)	0.19
Gate Valve (3/4 open)	1.0
Gate Valve (1/2 open)	5.6
Ball Check Valve (fully open)	70
Foot Valve (fully open)	15
Swing Check Valve (fully open)	2.3
Close Return Bend	2.2
Tee, Through Side Outlet	1.8
Standard Short Radius Elbow	0.9
Medium Sweep Elbow	0.8
Long Sweep Elbow	0.6
45° Elbow	0.4

or

$$(f(1500) + 10)\frac{Q^2}{5.62} = 30$$

Since f depends on the flow rate, a trial solution is needed. It might be begun by assuming $f = 0.02$, then $Q = 2.05$ cfs and $Re = 6.4 \times 10^5$. Now from the Moody diagram it can be determined that $f = 0.024$. Consequently,

$$(36 + 10)Q^2 = 168.6$$

or $Q = 1.91$ cfs. In this case the valve causes approximately 22 percent of the total head loss.

Entrances, Exits, Contractions, and Enlargements

When water which is stationary in a reservoir is accelerated radially inward in entering a pipe, separation will occur unless the inlet is sufficiently rounded. Consequently, the amount of head loss at a pipe entrance will depend upon the geometry of the entrance. The loss due to the entrance, however, can be evaluated by Eq. 3-1 with an appropriate coefficient K_L for the particular entrance. For a well rounded entrance the loss coefficient is $K_L = 0.05$; it is $K_L = 0.5$ for a square entrance; and if the pipe protrudes into the reservoir it is $K_L = 1.0$. These conditions are illustrated on the following page

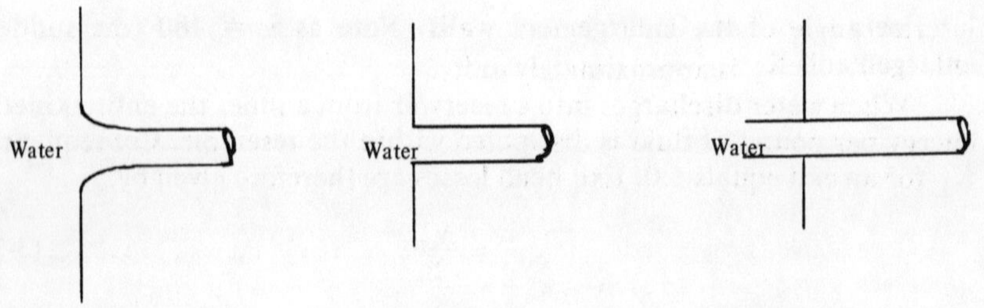

Rounded Entrance
$K_L = 0.05$

Square Entrance
$K_L = 0.5$

Re-entrant
Entrance
$K_L = 1.0$

A contraction behaves very similar to an entrance with the exception that the fluid need not accelerate radially inward as much. Since the loss is primarily due to separation in the pipe, head loss coefficients for pipe contractions are only slightly less than those due to entrances. Consequently for a *gradual* contraction $K_L = 0.04$, and for an *abrupt* contraction $K_L = 0.5$.

For enlargements a more appropriate equation for determining the minor head loss is,

$$h_L = K_L \frac{(V_1 - V_2)^2}{2g} \qquad \text{(3-2)}$$

in which V_1 is the average velocity in the smaller upstream pipe and V_2 is the average velocity in the larger downstream pipe. The form of this equation can be obtained from the momentum principle. This principle indicates that for a sudden enlargement $K_L = 1.0$. If the enlargement is gradual, Fig 3-3 gives appropriate values for K_L as a function of the

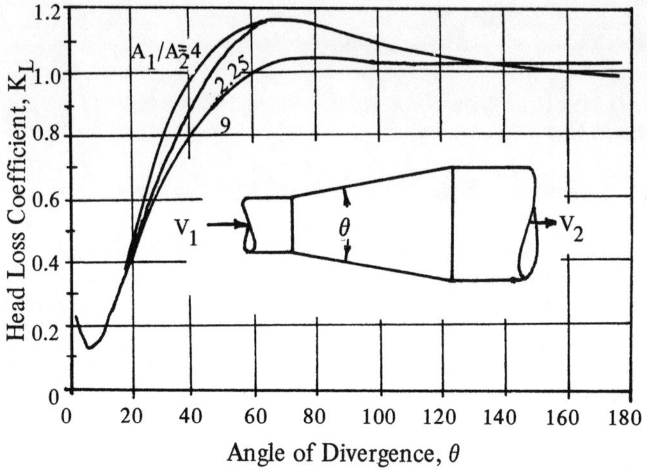

Fig. 3-3. Head loss coefficients for pipe enlargements.

interior angle of the enlargement walls. Note as $\theta = 180$ (the sudden enlargement) K_L is approximately unity.

When water discharges into a reservoir from a pipe, the entire kinetic energy per pound of fluid is dissipated within the reservoir. Consequently K_L for an exit equals 1.0. Exit head losses are therefore given by,

$$h_L = \frac{V^2}{2g} \qquad \text{(3-3)}$$

Chapter IV
Incompressible Flow in Pipe Networks

Introduction

Analyses and design of pipe networks create a relatively complex problems, particularly if the network consists of a large number of pipes as frequently occurs in the water distribution systems of large metropolitan areas, or natural gas pipe networks. Professional judgment is involved in deciding which pipes should be included in a single analysis. Obviously it is not practical to include all pipe which delivers to all customers of a large city, even though they are connected to the total delivery system. Often only those main trunk lines which carry the fluid between separate sections of the area are included, and if necessary analyses of the networks within these sections may be included. This manual deals only with steady-state solutions. In a water distribution system, the steady-state analysis is a small but vital component of assessing the adequacy of a network. Such an analysis is needed each time changing patterns of consumption or delivery are significant or add-on features, such as supplying new subdivisions, addition of booster pumps, or storage tanks change the system. In addition to steady analyses, studies dealing with unsteady flows or transient problems, operation and control, acquisition of supply, optimization of network performance against cost, and social implications should be given consideration but are beyond the scope of this text.

The steady-state problem is considered solved when the flow rate in each pipe is determined under some specified patterns of supply and consumption. The supply may be from reservoirs, storage tanks and/or pumps or specified as inflow or outflow at some point in the network. From the known flow rates the pressures or head losses throughout the system can be computed. Alternatively, the solution may be initially for the heads at each junction or node of the network and these can be used to compute the flow rates in each pipe of the network.

The oldest method for systematically solving the problem of steady flow in pipe networks is the Hardy Cross method. Not only is this method suited for hand solutions, but it has been programmed widely for computer solutions, but particularly as computers allowed much larger networks to be analyzed it become apparent that convergence of the Hardy Cross method might be very slow or even fail to provide a solution in some cases. In the past few years the Newton-Raphson method has been utilized to solve large networks, and with improvements in algorithms based on the Newton-Raphson method, computer storage requirements are not greatly larger than those needed by the Hardy Cross method. An additional method called the "linear theory method" has also been proposed, and does not require an initialization as do the other two methods but more storage. The solution of a system of algebraic equations gives the flow rates in each pipe or the head or pressures throughout the system described in this chapter. In the subsequent three chapters the implementation of the linear theory method, the Newton-Raphson method and the Hardy Cross method is discussed, in the reverse order of their historical development.

Reducing Complexity of Pipe Networks

In general, pipe networks may include series pipes, parallel pipes, and branching pipes (i.e. pipes that form the topology of a tree). In addition, elbows, valves, meters, and other devices which cause local disturbances and minor losses may exist in pipes. All of the above should be combined with or converted to an "equivalent pipe" in defining the network to be analyzed. The concept of equivalence is useful in simplifying networks. Methods for defining an equivalent pipe for each of the above mentioned occurrences follows.

Series pipes

The method for reducing two or more pipes of different size in series will be explained by reference to the diagram below. The same flow must

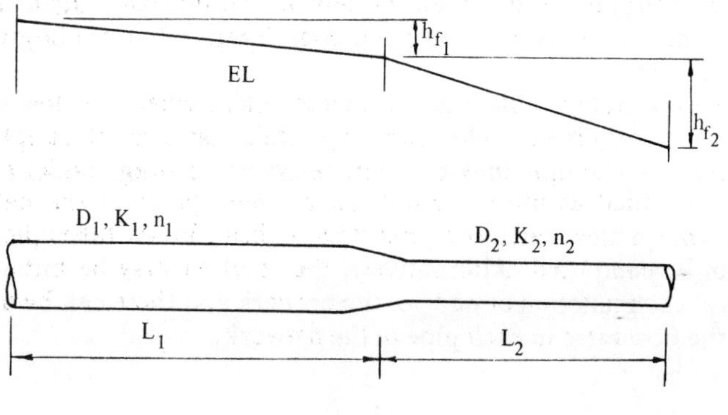

pass through each pipe in series. An equivalent pipe is a pipe which will carry this flow rate and produce the same head loss as two or more pipes, or

$$h_{f_e} = \Sigma h_{f_i} \qquad (4\text{-}1)$$

Expressing the individual head losses by the exponential formula gives,

$$K_e Q^{n_e} = K_1 Q^{n_1} + K_2 Q^{n_2} + \ldots = \Sigma K_i Q^{n_i} \qquad (4\text{-}2)$$

For network analysis K_e and n_e are needed to define the equivalent pipe's hydraulic properties. If the Hazen-Williams equation is used, all exponents $n = 1.852$, and consequently

$$K_e = K_1 + K_2 + \ldots = \Sigma K_i \qquad (4\text{-}3)$$

or the coefficient K_e for the equivalent pipe equals the sum of K's of the individual pipes in series. If the Darcy-Weisbach equation is used, the exponents n in Eq. 4-2 will not necessarily be equal, but generally these exponents are near enough equal that the n_e for the equivalent pipe can be taken as the average of these exponents and Eq. 4-3 used to compute K_e.

Parallel pipes

An equivalent pipe can also be used to replace two or more pipes in parallel. The head loss in each pipe between junctions where parallel pipes part and join again must be equal, or

$$h_f = h_{f_1} = h_{f_2} = h_{f_3} = \ldots \qquad (4\text{-}4)$$

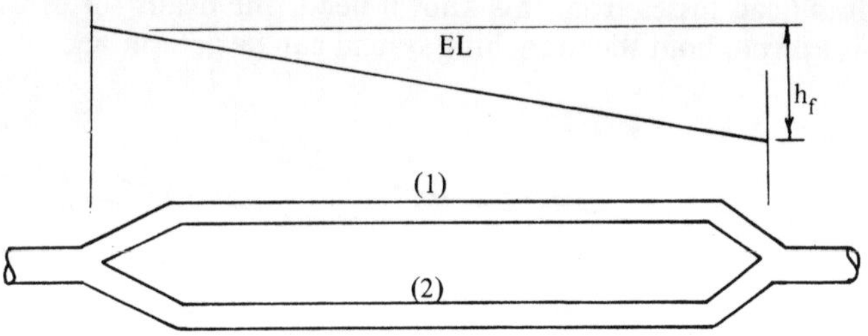

The total flow rate will equal the sum of the individual flow rates or

$$Q = Q_1 + Q_2 + \ldots = \Sigma Q_i \qquad (4\text{-}5)$$

Solving the exponential formula $h_f = KQ^n$ for Q and substituting into Eq. 4-5 gives

$$\left(\frac{h_f}{K_e}\right)^{\frac{1}{n_e}} = \left(\frac{h_f}{K_1}\right)^{\frac{1}{n_1}} + \left(\frac{h_f}{K_2}\right)^{\frac{1}{n_2}} + \ldots = \Sigma \left(\frac{h_f}{K_i}\right)^{\frac{1}{n_i}} \qquad \ldots\ldots (4\text{-}6)$$

If the exponents are equal as will be the case in using the Hazen-Williams equation, the head loss h_f may be eliminated from Eq. 4-6 giving

$$\left(\frac{1}{K_e}\right)^{\frac{1}{n}} = \left(\frac{1}{K_1}\right)^{\frac{1}{n}} + \left(\frac{1}{K_2}\right)^{\frac{1}{n}} + \ldots = \Sigma \left(\frac{1}{K_i}\right)^{\frac{1}{n}} \qquad \ldots\ldots\ldots\ldots (4\text{-}7)$$

When the Darcy-Weisbach equation is used for the analysis, it is common practice to assume n is equal for all pipes and use Eq. 4-7 to compute the K_e for the equivalent pipe.

Branching system

In a branching system a number of pipes are connected to the main to form the topology of a tree. Assuming that the flow is from the main into the smaller laterals it is possible to calculate the flow rate in any pipe as the sum of the downstream consumptions or demands. If the laterals supply water to the main, as in a manifold, the same might be done. In either case by proceeding from the outermost branches toward the main or "root of the tree" the flow rate can be calculated, and from the flow rate in each pipe the head loss can be determined using the Darcy-Weisbach or Hazen-Williams equation. In analyzing a pipe network containing a branching system, only the main is included with the total flow rate specified by summing from the smaller pipes. Upon completing the analysis the pressure head in the main will be known. By substracting individual head losses from this known head, the heads (or pressures) at any point throughout the branching system can be determined.

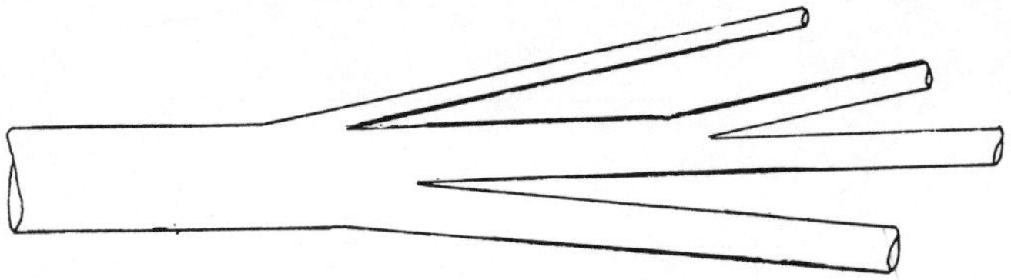

Minor losses

When a valve, meter, elbow, or other device exists in a pipe causing a minor loss which is not insignificant in comparison to the frictional loss in that pipe, an equivalent pipe should be formed for use in the network analyses. This equivalent pipe should have the same head loss for any flow rate as the sum of the pipe frictional loss and the minor head loss. The equivalent pipe is formed by adding a length ΔL to the actual pipe length such that the frictional head loss in the added length of pipe equals the minor losses. Computation of ΔL will be slightly different depending upon whether the Darcy-Weisbach or Hazen-Williams equations are to be used.

Most minor losses are computed from the formula Eq. 3-1 $h_L = K_L (V^2/2g)$ or a loss coefficient multiplied by the velocity head as described in Chapter III. The Darcy-Weisbach equation $h_f = [f(L/D)](V^2/2g)$ may also be thought of as the product of a coefficient times the velocity head. Consequently, if the Darcy-Weisbach equation is to be used in the network analyses, the length ΔL can be found by equating these two coefficients with ΔL replacing L in the Darcy-Weisbach coefficient. After solving for ΔL, the length to be added to the actual pipe length is,

$$\Delta L = \frac{K_L D}{f} \qquad (4\text{-}8)$$

Since f is generally a function of the flow rate, ΔL also depends upon the flow rate. In practice it is generally adequate to compute ΔL, by using the f values for wholly rough flow, or if knowledge of the approximate flow rate is available the friction factor f corresponding to it may be used in Eq. 4-8. If several devices causing minor losses exist in a single pipe, then the sum of the individual ΔL's is added to the length of the actual pipe.

The coefficient K in the exponential formula for the equivalent pipe is obtained by substituting L + ΣΔL for the length of the pipe in Eq. 2-28 or

$$K_e = \frac{a(L + \Sigma\Delta L)}{2g\,DA^2} \qquad (4\text{-}9)$$

when using the Darcy-Weisbach equation for computing frictional losses. When using the Hazen-Williams formula for this purpose the added pipe length ΔL, due to the device causing the minor loss, can be computed from

ES $\quad \Delta L = 0.00532\, K_L\, Q^{0.148}\, C_{HW}^{1.852}\, D^{0.8703}$

SI $\quad \Delta L = 0.00773\, K_L\, Q^{0.148}\, C_{HW}^{1.852}\, D^{0.8703}$

and the K in the exponential formula is

ES $\quad K_e = 4.73 \dfrac{L + \Delta L}{C_{HW}^{1.852}\, D^{4.87}}$ (D and L in feet)

SI $\quad K_e = \dfrac{10.7\,(L + \Delta L)}{C_{HW}^{1.852}\, D^{4.87}} \qquad (4\text{-}11)$

Example Problem in Finding Equivalent Pipes

1. A 12-inch (30.48 cm) cast iron pipe which is 100 ft (30.48 m) long is attached in series to a 10-inch (25.4 cm) dia. iron pipe which is 300 ft (91.44 m) long to carry a flow rate of 5.0 cfs (0.142 cm) of water at 68°F (20°C). Find the length of 10-inch (25.4 cm) pipe which is equivalent to the series system.

Solution:

ES

$$V_{12} = 5/0.785 = 6.37 \text{ fps}$$
$$V_{10} = 5/0.545 = 9.17 \text{ fps}$$
$$Re_{12} = 6.37(1)/1.084 \times 10^{-5} = 5.90 \times 10^{5}$$
$$Re_{10} = 9.17(0.833)/1.084 \times 10^{-5} = 7.05 \times 10^{5}$$
$$(e/D)_{12} = 0.0102/12 = 0.00085$$
$$(e/D)_{10} = 0.0102/10 = 0.00102$$

from the Moody diagram

$$f_{12} = 0.0195, f_{10} = 0.0192$$
$$h_f = h_{f_{12}} + h_{f_{10}}$$
$$h_f = 0.0195 \frac{100}{1} \frac{(6.37)^2}{64.4} + 0.0192 \frac{300}{0.833} \frac{(9.17)^2}{64.4}$$
$$h_f = 1.23 + 9.02 = 10.25 \text{ ft.}$$
$$L = h_f D(2g)/fV^2 = 340.7 \text{ ft.}$$

SI

$$V_{12} = 0.142/0.0729 = 1.95 \text{ m/s}$$
$$V_{10} = 0.142/0.0507 = 2.80 \text{ m/s}$$
$$Re_{12} = 1.95(0.3048)/1.007 \times 10^{-6} = 5.90 \times 10^{5}$$
$$Re_{10} = 2.80(0.254)/1.007 \times 10^{-6} = 7.05 \times 10^{5}$$
$$(e/D)_{12} = 0.0259/30.48 = 0.00085$$
$$(e/D)_{10} = 0.0259/25.4 = 0.00102$$

from the Moody diagram

$$h_f = 0.0195 \frac{30.48}{0.3048} \frac{(1.95)^2}{19.62} + 0.0192 \frac{91.44}{0.254} \frac{(2.80)^2}{19.62}$$
$$h_f = 0.378 + 2.762 = 3.14 \text{ m}$$
$$L = 3.14\,(0.254)\,(19.62)/(0.0192 \times 2.80^2) = 103.9 \text{ m}$$

2. **Using the Hazen-Williams formula find the coefficient K_e in the exponential formula and the diameter of an equivalent pipe to replace two 500 ft parallel pipes of 8-inch and 6-inch diameters. $C_{HW} = 120$ for both pipes, and make the equivalent pipe 500 ft long.**

Solution:

$$K_8 = \frac{4.73\ L}{C_{HW}^{1.852}\ D^{4.87}} = \frac{4.73\ (500)}{7090(0.667)^{4.87}} = 2.403$$

$$K_6 = \frac{4.73\ (500)}{7090(0.5)^{4.87}} = 9.754$$

From Eq. 4-7

$$\left(\frac{1}{K_e}\right)^{0.54} = \left(\frac{1}{2.403}\right)^{0.54} + \left(\frac{1}{9.754}\right)^{0.54} = 0.915$$

$$K_e = 1.178\ \text{ft}$$

$$D_e = \left[\frac{4.73\ L}{C_{HW}^{1.852}\ K_e}\right]^{1/4.87} = \left[\frac{4.73\ (500)}{7090\ (1.178)}\right]^{0.2053}$$

$$= 0.772\ \text{ft}$$

$$= 9.26\ \text{inches}$$

3. **An 800 ft long 8-inch cast iron pipe contains an open globe valve. Determine the length of the equivalent pipe if the flow rate is approximately 700 gpm.**

Solution:

Using the procedure for determining f described in Chapter II, f = 0.0218. From Eq. 4-8

$$\Delta L = \frac{K_L D}{f} = \frac{10(8/12)}{0.0218} = 306\ \text{ft}$$

$$L_e = L + \Delta L = 1106\ \text{ft}$$

Systems of Equations Describing Steady Flow in Pipe Networks

Flow rates as unknowns

The analysis of flow in networks of pipes is based on the continuity and energy laws as described in Chapter I. To satisfy continuity, the mass, weight, or volumetric flow rate into a junction must equal the mass, weight, or volumetric flow rate out of a junction. If the volumetric flow rate is used this principle, as discussed in Chapter I, can be expressed mathematically as,

$$(\Sigma Q_i)_{out} - (\Sigma Q_i)_{in} = C \quad \ldots\ldots\ldots\ldots\ldots (4\text{-}12)$$

in which C is the external flow at the junction (commonly called consumption or demand). C is positive if flow is into the junction and negative if it is out from the junction. For example if four pipes meet at a junction as shown in the sketch, Eq. 4-12 at this junction is

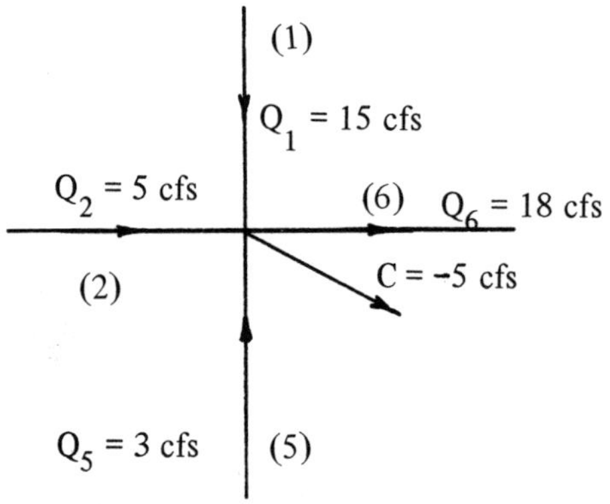

$$Q_6 - Q_1 - Q_2 - Q_5 = -5$$
$$18 - 15 - 5 - 3 = -5$$

If a pipe network contains J junctions (also called nodes) and all external flows are known then J-1 independent continuity equations in the form of Eq. 4-12 can be written. The last, or the Jth continuity equation, is not independent; that is, it can be obtained from some combination of the first J-1 equations. Note in passing that each of these continuity equations is linear, i.e., Q appears only to the first power.

In addition to the continuity equations which must be satisfied, the energy principle provides equations which must be satisfied. These additional equations are obtained by noting that if one adds the head losses around a closed loop, taking into account whether the head loss is positive or negative, that upon arriving at the beginning point the net head losses equals zero. Mathematically, the energy principle gives L equations of the form

$$\sum_{\ell}^{I} h_{f_\ell} = 0$$

$$\sum_{\ell}^{II} h_{f_\ell} = 0 \quad \text{(4-13)}$$

$$\vdots$$

$$\sum_{\ell}^{L} h_{f_\ell} = 0$$

in which L represents the number of non-overlapping loops (also referred to as natural loops) in the network, and the summation on small ℓ is over the pipes in the loops I, II, ..., L. By use of the exponential formula $h_f = KQ^n$, Eqs. 4-13 can be written in terms of the flow rate, or

$$\sum_{\ell}^{I} K_\ell Q_\ell^{n_\ell} = 0$$

$$\sum_{\ell}^{II} K_\ell Q_\ell^{n_\ell} = 0 \quad \text{(4-14)}$$

$$\vdots$$

$$\sum_{\ell}^{L} K_\ell Q_\ell^{n_\ell} = 0$$

A pipe network consisting of J junctions and L non-overlapping loops and N pipes will satisfy the equation

$$N = (J - 1) + L \quad \text{(4-15)}$$

(If all of the external flows are not known, then all J junction equations are independent and available for use as will be discussed in the next chapter.) Since the flow rate in each pipe can be considered unknown, there will be N unknowns. The number of independent equations which can be obtained for a network as described above are (J-1) + L. Consequently the number of independent equations is equal in number to the unknown flow rates in the N pipes. The (J-1) continuity equations are linear and the L energy (or head losses) equations are nonlinear. Since large networks may consist of hundreds of pipes, systematic methods which utilize computers are needed for solving this system of simultaneous equations. Such methods are described in subsequent chapters.

As an example in defining the system of N equations which must be satisfied in solving for the N unknown volumetric flow rates in the N pipes of a network, consider the simple two loop network given below. In this

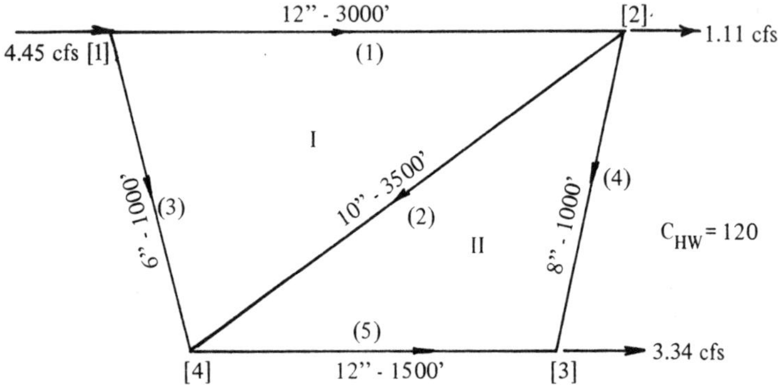

example there are five pipes and therefore five unknown flow rates. There are four junctions and therefore three independent continuity equations and two energy equations for the head losses around the two basic loops can be written. On the sketch of this network the pipe numbers are enclosed by parentheses, the junction or node numbers are within [ ] brackets and the loops are denoted by Roman Numerals I and II. Arrow heads denote assumed directions of flows in the pipes. Flow rates, etc., will be denoted by a subscript corresponding to the pipe number in which that flow rate occurs. This same notation will be followed throughout the remainder of the manual. Also considerations of space prevent duplicating solutions in ES and SI units.

The J-1 = 3 continuity equations at the three consecutive junctions 1, 2, and 3 are

$$\begin{aligned} Q_1 + Q_3 \quad &= 4.45 \\ -Q_1 + Q_2 + Q_4 &= -1.11 \\ -Q_4 - Q_5 \quad &= -3.34 \end{aligned}$$

The continuity equation at junction 4 is $-Q_3 - Q_2 + Q_5 = 0$. However, this equation is not independent of the above three equations since it can be obtained as minus the sum of these three equations. The Hazen-Williams equation will be used to define the head losses in each pipe. In expressing these head-losses the exponential equation will be used. From Eq. 2-22 the K coefficients for the exponential formula are: $[K = 4.73L/(C_{HW}^{1.852} D^{4.87})]$:

$$K_1 = 2.018, K_2 = 5.722, K_3 = 19.674, K_4 = 4.847, K_5 = 1.009$$

The energy loss equations around the two loops are:

$$2.018\, Q_1^{1.852} + 5.722\, Q_2^{1.852} - 19.674\, Q_3^{1.852} = 0$$

$$4.847\, Q_4^{1.852} - 1.009\, Q_5^{1.852} - 5.722\, Q_2^{1.852} = 0$$

These two energy equations are obtained by starting at junctions 1 and 2, respectively, and moving around the respective loops I and II in a clockwise direction. If the assumed direction of flow is opposite to this clockwise movement a minus precedes the head loss for that pipe. Simultaneous equations such as those above will be called *Q-equations.*

A solution to the five unknown flow rates from the five simultaneous equations above, by the procedure described subsequently in Chapter V is: $Q_1 = 3.350$ cfs, $Q_2 = 0.897$ cfs, $Q_3 = 1.104$ cfs, $Q_4 = 1.340$ cfs, $Q_5 = 2.001$ cfs. This solution can be verified by substituting into each of the above equations. It is relatively easy to determine flows in individual pipes which also satisfy the J-1 continuity equations. However, the correct flow rates which simultaneously satisfy the L energy equations are virtually impossible to obtain by trial and error if the system is large.

After solving the system of equations for the flow rate in each pipe, the head losses in each pipe can be determined. From a known head or pressure at one junction it is then a routine computation to determine the heads and pressures at each junction throughout the network, or at any point along a pipe, by substracting the head loss from the head at the upstream junction, plus accounting for differences in elevations if this be the case. In some problems the external flows may not be known as was assumed in the above example. Rather, the supply of water may be from reservoirs and/or pumps. The amount of flow from these individual sources will not only depend upon the demands, but also will depend upon the head losses throughout the system. Methods for incorporating pumps and reservoirs into a network analysis in which the flow rates in the individual pipes of the network are initially considered the unknowns will be dealt with in Chaper V in conjunction with the linear theory method of solution.

Example Problem in Writing Flow Rate Equations

1. Write the system of equations whose solution provides the flow rates in the six pipes of the network shown below. The energy equations are to be based on the Darcy-Weisbach equation.

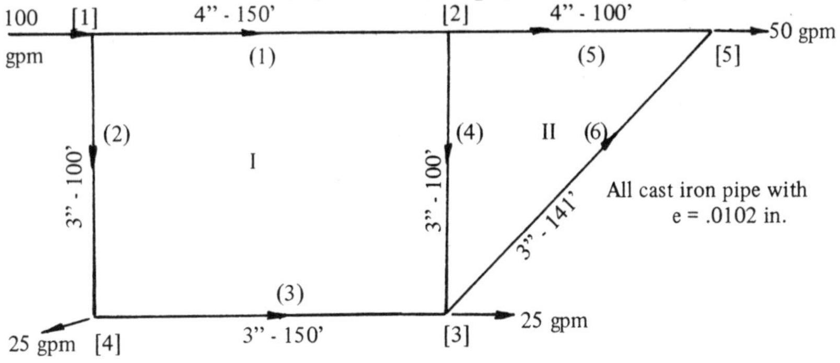

Solution:

Before the energy equations can be defined it is necessary that K and n for each pipe be determined for the range of flow rates expected in that pipe by the procedures described in Chapter II. This might be accomplished by a computer program which determines f for the specified flow rate plus an incremental flow rate and f for the specified flow rate minus an incremental flow rate, and from these f's and Q's compute a and b in Eq. 2-26 and thereafter K and n for the exponential formula. In solving a pipe network problem a computer algorithm for doing this might be called upon whenever the flow rates being used in the solution are outside the range for which K and n are applicable. A listing of a FORTRAN program for accomplishing these computations follows along with the input data required for this problem. Values for K and n for each pipe is given below the listing.

```
100 FORMAT(8F10.5)
    ELOG=9.35*ALOG10(2.71828183)
 20 READ(5,100,END=99) Q,D,FL,E,DQP,VIS
    DEQ=Q*DQP
    ED=E/D
    D=D/12.
    A=.78539392*D**2
    Q1=Q-DEQ
    Q2=Q+DEQ
    V1=Q1/A
    V2=Q2/A
    RE1=V1*D/VIS
    RE2=V2*D/VIS
    ARL=FL/(64.4*D*A**2)
    F=1./(1.14-2.*ALOG10(ED))**2
    RE=RE1
    MM=0
 57 MCT=0
 52 FS=SQRT(F)
    FZ=.5/(F*FS)
    ARG=ED+9.35/(RE*FS)
    FF=1./FS-1.14+2.*ALOG10(ARG)
    DF=FZ+ELOG*FZ/(ARG*RE)
    DIF=FF/DF
    F=F+DIF
    MCT=MCT+1
    IF(ABS(DIF) .GT. .00001 .AND. MCT .LT. 15) GO TO 52
    IF(MM .EQ. 1) GO TO 55
    MM=1
    RE=RE2
    F1=F
    GO TO 57
 55 F2=F
    BE=(ALOG(F1)-ALOG(F2))/(ALOG(Q2)-ALOG(Q1))
    AE=F1*(Q-DEQ)**BE
    EP=2.-BE
    CK=AE*ARL
    WRITE(6,101) Q,D,BE,AE,EP,CK
```

```
101 FORMAT(1H ,5F12.5,E16.6)
    GO TO 20
 99 STOP
    END
```

Input data for above problem.

Q	D	L	e	ΔQ ratio	ν
0.12	4.	150.	0.0102	0.1	0.00001217
0.10	3.	100.	0.0102	0.1	0.00001217
0.05	3.	150.	0.0102	0.1	0.00001217
0.05	3.	100.	0.0102	0.1	0.00001217
0.1	4.	100.	0.0102	0.1	0.00001217
0.05	3.	141.	0.0102	0.1	0.00001217

Pipe No.	1	2	3	4	5	6
Q(cfs)	0.12	0.10	0.05	0.05	0.10	0.05
K	21.0	63.9	85.2	56.8	13.6	80.1
n	1.90	1.92	1.88	1.88	1.89	1.88

The equations which will provide a solution are:

$$Q_1 + Q_2 = 0.223$$

$$-Q_1 + Q_4 + Q_5 = 0$$

$$-Q_3 - Q_4 + Q_6 = -0.056$$

$$-Q_2 + Q_3 = -0.056$$

$$21.0\,Q_1^{1.90} + 56.8\,Q_4^{1.88} - 85.2\,Q_3^{1.88} - 63.9\,Q_2^{1.92} = 0$$

$$13.6\,Q_5^{1.89} - 80.1\,Q_6^{1.88} - 56.8\,Q_4^{1.88} = 0$$

Heads at Junctions as Unknowns

If the head (either the total head or the piezometric head, since the velocity head is generally ignored in determining heads or pressure in pipe networks) at each junction is initially considered unknown instead of the flow rate in each pipe, the number of simultaneous equations which must be solved can be reduced in number. The reduction in number of equations, however, is at the expense of not having some linear equations in the system.

To obtain the system of equations which contain the heads at the junctions of the network as unknowns, the J-1 independent continuity equations are written as before. Thereafter the relationship between the flow rate and head loss is substituted into the continuity equations. In writing these equations it is convenient to use a double subscript for the flow rates. These subscripts correspond to the junctions at the ends of the pipe. The first subscript is the junction number from which the flow comes and the second is the junction number to which the flow is going. Thus Q_{12} represents the flow in the pipe connecting junctions 1 and 2 assuming the flow is from junction 1 to junction 2. If the flow is actually in this direction Q_{12} is positive and Q_{21} equals minus Q_{12}. Solving for Q from the exponential formula (using the double subscript notation) gives

$$Q_{ij} = (h_{L_{ij}}/K_{ij})^{1/n_{ij}} = \left(\frac{H_i - H_j}{K_{ij}}\right)^{1/n_{ij}} \quad \text{.........(4-16)}$$

If Eq. 4-16 is substituted into the junction continuity equations (Eq. 4-12), the following equation results:

$$\left[\Sigma\left(\frac{H_i - H_j}{K_{ij}}\right)^{\frac{1}{n_{ij}}}\right]_{out} - \left[\Sigma\left(\frac{H_i - H_j}{K_{ij}}\right)^{\frac{1}{n_{ij}}}\right]_{in} = C \quad \text{........(4-17)}$$

Upon writing an equation of the form of Eq. 4-17 at J-1 junctions, a system of J-1 nonlinear equations is produced.

As an illustration of this system of equations with the heads at the junctions as the unknowns, consider the simple one loop network below which consists of three junctions and three pipes. In this network two independent continuity equations are available and consequently the head at one of the junctions must be known. In this case at [1]. The two continuity equations are:

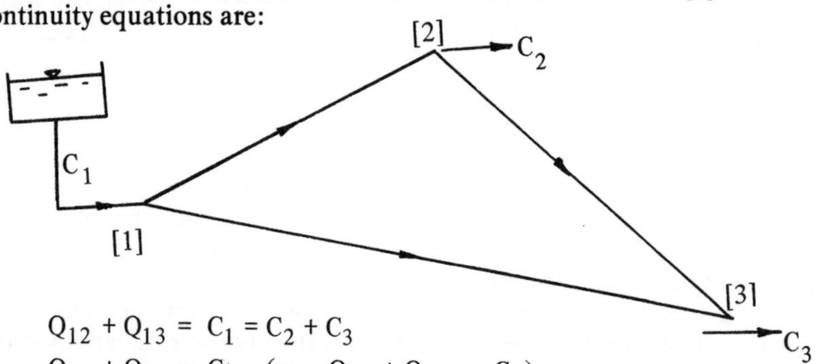

$$Q_{12} + Q_{13} = C_1 = C_2 + C_3$$

$$Q_{21} + Q_{23} = -C_2 \quad (\text{or}, -Q_{12} + Q_{23} = -C_2)$$

Note that even though in the second equation the flow in pipe 1-2 is toward the junction, the flow rate Q_{21} is not preceded by a minus sign since the notation 2-1 takes care of this. Alternatively the equations could have been written at junctions 2 and 3 instead of 1 and 2. Substituting Eq. 4-16 into these continuity equations gives the following two equations to solve for the heads, H_2 and H_3 (H_1 is known and the subscripts of the H's denote the junction numbers):

$$\left(\frac{H_1 - H_2}{K_{12}}\right)^{1/n_{12}} + \left(\frac{H_1 - H_3}{K_{13}}\right)^{1/n_{13}} = C_2 + C_3$$

$$-\left(\frac{H_1 - H_2}{K_{12}}\right)^{1/n_{12}} + \left(\frac{H_2 - H_3}{K_{23}}\right)^{1/n_{13}} = -C_2$$

Since a negative number cannot be raised to a power a minus sign must precede any term in which the subscript notation is opposite to the direction of flow, i.e. the second form of equation as given in parentheses is used. Systems of these equations will be referred to as H-equations.

Upon solving this nonlinear system of equations, the pressure at any junction j can be computed by substracting the junction elevation from the head h_j and then multiplying this difference by γ the specific weight of the fluid. To determine the flow rates in the pipes of the network, the now known heads are substituted into Eq. 4-16.

Corrective flow rates around loops of network considered unknowns

Since the number of junctions minus 1 (i.e. J-1) will be less in number than the number of pipes in a network by the number of loops L in the network, the last set of H-equations will generally be less in number than the system of Q-equations. This reduction in number of equations is not necessarily an advantage since all of the equations are nonlinear, whereas in the system of Q-equations only the L energy equations were nonlinear. A system which generally consists of even fewer equations can be written for solving a pipe network, however. These equations consider a corrective flow rate in each loop as the unknowns. This latter system will be referred to as the ΔQ-equations. Since there are L basic loops in a network the ΔQ-equations consist of L equations, all of which are nonlinear.

It is not difficult to establish an initial flow in each pipe which satisfies the J-1 junction continuity equation (which must also satisfy the Jth junction continuity equation). These initial flow estimates generally will not simultaneously satisfy the L head loss equations. Therefore they must be corrected before they equal the true flow rates in the pipes. A flow rate adjustment can be added (accounting for sign) to the initially assumed flow in each pipe forming a loop of the network without violating continuity at the junctions. This fact suggests establishing L energy (or head loss) equations around the L loops of the network in which the initial flow plus the corrective loop flow rate ΔQ is used as the true flow rate in the head loss equations. Upon satisfying these head loss equations by finding the appropriate corrective loop flow rates, the J-1 continuity equations would remain satisfied as they initially were. The corrective loop flow rates ΔQ_i may be taken positive in the clockwise or counterclockwise direction around the basic loops, but the sign convention must be consistent around any particular loop, and preferably in the same direction of all loops of the network. Clockwise directions will be considered positive in this book.

In order to develop the mathematics of this possible system of ΔQ-equations, the initially assumed flows, which satisfy the junction continuity equations, will contain an o subscript as well as an i to denote the pipe number. Thus Q_{oi}, i = 1,2 ... N represents the initially assumed flow rates in the N pipes. The corrective loop flow rates will be denoted by ΔQ_ℓ. Thus ΔQ_ℓ, ℓ = 1,2 ... L are corrective flow rates around the L loops of the system which must be added to Q_o for a given pipe to get the actual flow rate in that pipe. Using this notatin the L energy equations around the basic loops can be written as,

$$\sum_i^{I} K_i (Q_{o_i} + \overrightarrow{\Delta Q}_1)^{n_i} = 0 \quad \text{(head loss around loop I)}$$

$$\sum_i^{II} K_i (Q_{o_i} + \overrightarrow{\Delta Q}_2)^{n_i} = 0 \quad \text{(head loss around loop II)}$$

$$\vdots \qquad \qquad \qquad \qquad \qquad \qquad (4\text{-}18)$$

$$\sum_i^{L} K_i (Q_{o_i} + \overrightarrow{\Delta Q}_L)^{n_i} = 0 \quad \text{(head loss around loop L)}$$

in which each summation includes only those pipes in the loop designated by the Roman numeral I, II, ... L, and $\overrightarrow{\Delta Q_\ell}$ always includes ΔQ_ℓ and also any other ΔQ's flowing through the pipe for which the term applies.

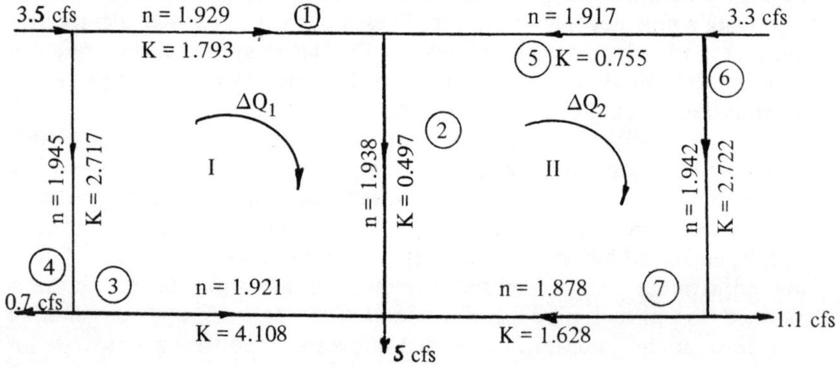

The system of equations, Eq. 4-18, will be set up for the two-loop network shown below. Values for K and n in the exponential formula for the expected flow rates are given by each pipe in the network. The two corrective loop flow rates ΔQ_1 and ΔQ_2 are taken as positive in the clockwise direction. The first step is to provide initial estimates of the flow rate in each pipe which satisfy the junction continuity equations. The estimates are: $Q_{01} = 1.75$ cfs, $Q_{02} = 3.55$ cfs, $Q_{03} = 1.05$ cfs, $Q_{04} = 1.75$ cfs, $Q_{05} = 1.8$ cfs, $Q_{06} = 1.5$ cfs, $Q_{07} = 0.4$ cfs in the directions shown by the arrows on the sketch. The head loss equations around the two loops are:

$$F_1 = 1.793\,(1.75 + \Delta Q_1)^{1.929} + 0.497\,(3.55 + \Delta Q_1 - \Delta Q_2)^{1.938} - 4.108\,(1.05 - \Delta Q_1)^{1.921} - 2.717(1.75 - \Delta Q_1)^{1.945} = 0$$

$$F_2 = -0.755\,(1.8 - \Delta Q_2)^{1.917} + 2.722\,(1.5 + \Delta Q_2)^{1.942} + 1.628\,(0.4 + \Delta Q_2)^{1.878} - 0.497\,(3.55 - \Delta Q_2 + \Delta Q_1)^{1.938} = 0$$

Upon obtaining the solution to these two equations for the two unknowns ΔQ_1 and ΔQ_2, the flow rates in each pipe can easily be determined by adding these corrective loop flow rates to the initially assumed flow rates. From these flow rates the head losses in each pipe are determined.

The nonlinearities in this system of equations, as well as the previous two systems discussed, make solution difficult. In the next three chapters methods for obtaining solutions are discussed. The Newton method and the Hardy Cross method (which is the Newton method applied to one equation at a time) are well adapted for the corrective loop flow rate equations, and also the junction head equations. These methods are described respectively in Chapters VI and VII. The Q-equations or the equations which consider the flow in each pipe unknown, can be solved by the linear theory method as discussed in the next chapter, Chapter V. This flow rate system of equations can be solved by the Newton method also. In fact the linear theory method is a variation of the Newton method.

Chapter V
Linear Theory Method

Introduction

In Chapter IV three alternative systems of equations were described which might be used in solving for the flow or pressure distribution in pipe networks. In this chapter the linear theory method[1] will be described and used in solving the system of equations which considers the flow rates unknown (i.e. the Q-equations). This system of equations is easy to use if the network is flow rate oriented, i.e. all external flows to the system are known. The linear theory method will be described first for solving the system of Q-equations. Thereafter it will be extended to networks containing pumps and reservoirs. For such networks, not all external flows are known, and must be obtained as part of the solution.

The linear theory method has several distinct advantages over the Newton-Raphson or the Hardy Cross methods described in the next two chapters. First, it does not require an initialization, and secondly according to Wood and Charles it always converges in a relatively few iterations. Its use in solving the head oriented equations or the corrective loop oriented equations is not recommended.

Transforming Nonlinear Energy Equations Into Linear Equations

Linear theory transforms the L nonlinear loop equations into linear equations by approximating the head in each pipe by,

$$h_{f_i} = \left[K_i\, Q_i\,(0)^{n-1}\right] Q_i = K_i' Q_i \qquad (5\text{-}1)$$

---

[1]Wood, Don J., and Carl O. A. Charles. "Hydraulic Network Analysis Using Linear Theory," Jour. of the Hydraulics Div., ASCE, 98(HY7):1157-1170. July 1972.

That is a coefficient K' is defined for each pipe as the produce of K_i multiplied by $Q(0)_i^{n-1}$, an estimate of the flow rate in that pipe. Combining these artificial linear loop equations with the J-1 junction continuity equations provide a system of N linear equations which can be solved by linear algebra. The solution will not necessarily be correct because of $Q_i(0)$'s will probably not have been estimated equal to the Q_i's produced by the solution. By repeating the process, after improving the estimates to $Q_i(1)$, eventually the $Q_i(m)$'s will equal the Q_i's. After this iteration the correct solution has been obtained.

In applying the linear theory method it is not necessary to supply an initial guess, as perhaps implied. Instead for the first iteration each K_i' is set equal to K_i, which is equivalent to setting all flow rates $Q_i(0)$ equal to unity. Wood, in developing the linear theory method, observed that successive iterative solutions tend to oscillate about the final solution.[2] He, therefore, suggests that after two iterative solutions have been obtained, and thereafter, that each flow rate used in the computations be the average flow rate for that pipe from the past two solutions or

$$Q_i(n) = \frac{Q_i(n\text{-}1) + Q_i(n\text{-}2)}{2} \qquad \text{(5-2)}$$

An example will be followed through in detail to clarify the process in solving a pipe network by the linear theory method. This example is the two-loop network shown below which consists of seven pipes. Pipes 1 and 2 contain globe valves and pipe 2 contains an orifice meter.

The first step is to obtain K and n for the exponential formula for a range of flow rates believed to be realistic. (In implementing the method in a computer program the first values used for K and n may be obtained from the Hazen-Williams equation.) Estimates of n and K are given in the table below the sketch of the pipe network (following page).

The second step is to form equivalent pipes for those pipes containing the globe valve and the orifice meter by use of Eq. 4-8,

$$\Delta L = \frac{K_L D}{f} = \frac{10(0.667)}{0.0218} = 306 \text{ ft} \quad \text{for globe valve in pipe (1)}$$

$$\Delta L = \frac{10(0.833)}{0.02085} = 400 \text{ ft} \quad \text{for globe valve in pipe (5)}$$

$$\Delta L = \frac{1.2(1)}{0.0238} = 51 \text{ ft} \quad \text{for the orifice meter in pipe (2)}$$

---

[2]Reasons for the oscillation can be understood by observing that the linear theory method is a variation of the Newton-Raphson method described in Chapter VI in which K_i would equal the derivative of h_f if multiplied by n_i. Thus the matrix of the linear theory method closely resembles the Jacobian of the Newton-Raphson method. The oscillation would be prevented by multiplying each K by its n but this would require more arithmetic than averaging consecutive solutions as proposed by Wood.

These lengths are added to the pipe lengths to form equivalent pipes. The K's in the table below the network are for the equivalent pipes.

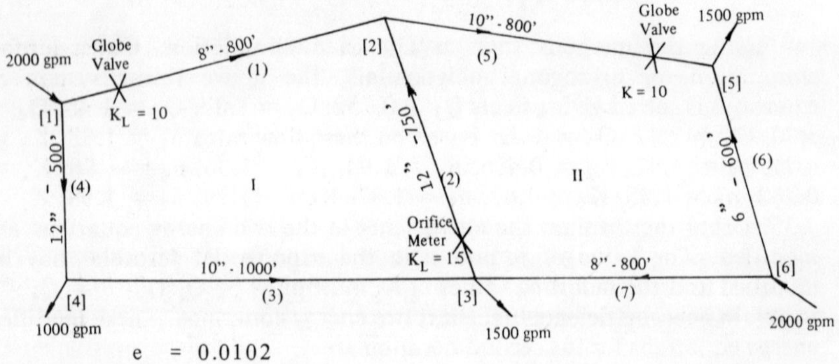

e = 0.0102

Pipe No.	1	2	3	4	5	6	7
V_1 (fps)	3.35	0.425	0.922	2.76	2.76	5.10	6.70
V_2 (fps)	6.70	0.851	1.84	5.53	5.51	10.21	13.40
$Re_1 \times 10^{-5}$	1.83	3.50	6.29	2.27	1.89	2.10	3.67
$Re_2 \times 10^{-5}$	3.67	6.99	1.26	4.54	3.77	4.19	7.34
f_1	0.0221	0.0250	0.0234	0.0203	0.0212	0.0234	0.0215
f_2	0.0215	0.0226	0.0218	0.0196	0.0205	0.0229	0.0212
b	0.0408	0.146	0.101	0.0472	0.0475	0.0287	0.0215
a	0.0223	0.0213	0.0218	0.0210	0.0216	0.0234	0.0219
n	1.96	1.85	1.90	1.95	1.95	1.97	1.98
K	4.71	0.402	1.37	0.264	1.14	11.30	3.35

The network has six junctions and therefore five independent continuity equations can be written. These are:

$$Q_1 + Q_4 = 4.45$$
$$-Q_1 - Q_2 + Q_5 = 0$$
$$Q_2 - Q_3 - Q_7 = -3.34$$
$$Q_3 - Q_4 = -2.23$$
$$-Q_5 - Q_6 = -3.34$$

A minus sign precedes those Q's with subscripts of pipes whose assumed directions of flow are into the junction, accoding to the convention adopted in Eq. 4-12. Under this convention external flows into the junction are positive and external flows, or demands which extract flow from the junction are negative. Alternatively, Q's for pipes whose flows are assumed into the junction may be taken positive and those out from the junction taken negative provided external demands are positive and incoming supplies are negative on the right of the equal sign. The two head loss equations are linearized by forming a coefficient for each Q_i which equals K multiplied by $Q_i(0)^{0.85}$. For the first iteration each $Q_i(0) = 1$ and therefore the linearized energy equations for the first iteration are:

$$4.71\,Q_1 - 0.402\,Q_2 - 1.37\,Q_3 - 0.264\,Q_4 = 0$$
$$1.14\,Q_5 - 11.30\,Q_6 + 3.35\,Q_7 + 0.402\,Q_2 = 0$$

Using the methods such as Gaussian elimination, Gauss-Jordan elimination, or orthogonal polynomials, the above linear system of equations is solved giving in cfs $Q_1 = 0.752$, $Q_2 = 1.37$, $Q_3 = 1.48$, $Q_4 = 3.70$, $Q_5 = 2.12$, $Q_6 = 1.22$. Based on these flow rates $n_1 = 1.93$, $K_1 = 4.71$, $n_2 = 1.92$, $K_2 = 0.406$, $n_3 = 1.94$, $K_3 = 1.36$, $n_4 = 1.96$, $K_4 = 0.262$. $n_5 = 1.95$, $K_5 = 1.62$, $n_6 = 1.97$, $K_6 = 11.29$, $n_7 = 1.98$, $K_7 = 3.35$. Using these values the coefficients in the two energy equations are modified. The value of K and n in the exponential formula may be modified and the modified values of K_i multiplied by $Q_i(1)^{ni-1} = Q_i^{ni-1}$ to obtain new coefficients in the last two energy equations. These modified energy equations for the second iteration are:

$$3.62\,Q_1 - 0.541\,Q_2 - 1.96\,Q_3 - 0.92\,Q_4 = 0$$
$$3.31\,Q_5 - 13.70\,Q_6 + 10.55\,Q_7 + 0.406\,Q_2 = 0$$

After solving this system, $Q_1 = 1.02$, $Q_2 = 0.653$, $Q_3 = 1.20$, $Q_4 = 3.43$, $Q_5 = 1.68$, $Q_6 = 1.66$, $Q_7 = 2.79$ cfs. Using the average of these flow rates and those from the previous solution to define the coefficients in the energy equations produces the final solution since these values do not change in the third digit beyond the decimal point during the next iteration. The solution therefore is: $Q_1 = 1.032$, $Q_2 = 0.656$, $Q_3 = 1.195$, $Q_4 = 3.422$, $Q_5 = 1.688$, $Q_6 = 1.653$, $Q_7 = 2.802$ cfs.

For a simple network such as that above it is practical to carry out the solution by hand. Obviously, for larger networks involving 100 or more pipes, computer algorithms must be written to carry out the numerous calculations. Below a FORTRAN computer program is listed for accomplishing this task. With this program the problem described above took 0.20 second of execution time on a UNIVAC 1108 computer. Description of the data input required by an expanded version of this program which allows for pumps and reservoirs is given in Appendix C. A program of just over one-half the size of the one listed can be written if the Hazen-Williams equation is used instead of the Darcy-Weisbach equation, since the n's and K's in the exponential formula do not depend upon the flow rate.

To use this program four different types of information are required. First the number of pipes, the number of junctions, and number of loops in the network plus other specifications such as denoting type of data (i.e. ES or SI), number of allowable iterations, viscosity of fluid, etc. (Card No. 1). Second, data giving the diameters, lengths, and wall roughness for each of the pipes (next three read statements). Third, information for establishing junction continuity equations. This information is provided by a card for each junction which contains the pipe numbers meeting at the junction. If the assumed direction of flow is into the junction this pipe number is preceded by a minus. Also if external flow occurs at the junction another card contains this data. Fourth, information is required for the energy equations. For each loop in the network a card lists the pipe numbers in that loop. A minus proceeds the pipe number if the assumed direction of flow is counterclockwise around the loop.

```
      INTEGER JN(40,5),NN(40),JB(20),IFLOW(40),LP(8,20), JC(50)
      REAL D(50),L(50),A(50,51),QJ(20),E(50),KP(50),V(2  ), Q(50),EXPP(50),
     $ AR(50),ARL(50)
   30 READ(5,110,END=99) NP,NJ, NL,MAX,NUNIT,ERR,VIS,DELQ1
  110 FORMAT(5I5,3F10.5)
C  NP--NO. OF PIPES, NJ--NO. OF JUNCTIONS, NL--NO. OF
C  LOOPS, MAX--NO. OF ITERATIONS ALLOWED, IF NUNIT=0
C  D AND E IN INCHES AND L IN FEET,IF NUNIT=1--D AND
C  E IN FEET AND L IN FEET, IF NUNIT=2 D AND E IN
C  METERS AND L IN METERS, IF NUNIT=3 D AND E IN CM
C  AND L IN METERS.
  100 FORMAT(16I5)
      NPP=NP+1
      NJ1=NJ-1
      READ(5,101) (D(I),I=1,NP)
      READ(5,101) (L(I),I=1,NP)
      READ(5,101) (E(I),I=1,NP)
  101 FORMAT(8F10.5)
      DO 48 I=1,NP
   48 E(I)=E(I)/D(I)
      IF (NUNIT-1) 40,41,42
   40 WRITE(6,102) (D(I),I=1,NP)
  102 FORMAT('OPIPE DIAMETERS (INCHES)',/,(1H ,16F8.1))
      DO 43 I=1,  NP)
   43 D(I)=D(I)/12.
      GO TO 44
   41 WRITE(6,112) (D(I),I=1,NP)
  112 FORMAT( 'OPIPE DIAMETERS (FEET)',/,(1H ,16F8.3))
   44 WRITE (6,103) (L(I),I=1,NP)
  103 FORMAT('OLENGTHS OF PIPE (FEET)',/,(1H ,16F8.0))
      G2=64.4
      GO TO 50
   42 IF (NUNIT .EQ. 2) GO TO 45
      DO 46 I=1,NP
   46 D(I)=.01*D(I)
   45 WRITE(6,113) (D(I),I=1,NP)
  113 FORMAT('O PIPE DIAMETERS (METERS)',/,(1H ,16F8.4))
      WRITE(6,114) (L(I),I=1,NP)
  114 FORMAT('O LENGTH OF PIPES (METERS)',/,(1H ,16F8.1))
      G2=19.62
      WRITE(6,115) (E(I),I=1,NP)
  115 FORMAT('O RELATIVE ROUGHNESS OF PIPES',/, (1H ,16F8.6))
C  INFLOW--IF 0 NO INFLOW, IF 1 THEN NEXT CARD GIVES
C  MAGNITUDE IN GPM. IF 2 NEXT CARD GIVES MAGNITUDE
C  IN CFS. IF 3 NEXT CARD GIVES MAGNITUDE IN CMS.
C  NNJ--NO. OF PIPES AT JUNCTIONS--POSITIVE FOR INFLOW
C  NEGATIVE FOR OUTFLOW. JN--THE NUMBERS OF PIPES
C  AT JUNCTION, IF FLOW ENTERS MINUS- IF FLOW LEAVES
C  THE PIPE NUMBER IS POSITIVE.
```

```
      DO 70 I=1,NP
      AR(I)=.78539392*D(I)**2
   70 ARL(I)=L(I)/(G2*D(I)*AR(I)**2)
      II=1
      DO 1 I=1,NJ
      READ(5,100) IFLOW(I),NNJ,(JN(I,J),J=1,NNJ)
      NN(I)=NNJ
      IF(IFLOW(I) - 1) 1,2,3
    2 READ(5,101) QJ(II)
      QJ(II)=QJ(II)/449.
      JB(II)=I
      GO TO 4
    3 READ(5,101) QJ(II)
      BJ(II)=I
    4 II=II+1
    1 CONTINUE
C   NUMBER OF PIPES IN EACH LOOP (SIGN INCLUDED)
      DO 35 I=1,NL
      READ(5,100) NNJ,(LP(J,I),J=1,NNJ)
   35 LP(8,I)=NNJ
      DO 5 I=1,NP
      IF (NUNIT .GT. 1) GO TO 66
      KP(I)=.0009517*L(I)/D(I)**4.87
      GO TO 5
   66 KP(I)=.00212*L(I)/D(I)**4.87
    5 CONTINUE
      ELOG=9.35*ALOG10(2.71828183)
      SUM=100.
      NCT=0
   20 II=1
      DO 6 I=1,NJ1
      DO 7 J=1,NP
    7 A(I,J)=0.
      NNJ=NN(I)
      DO 8 J=1,NNJ
      IJ=JN(I,J)
      IF(IJ .GT. 0) GO TO 9
      IIJ=ABS(IJ)
      A(I,IIJ)=-1.
      GO TO 8
    9 A(I,IJ)=1.
    8 CONTINUE
      IF(IFLOW(I) .EQ. 0) GO TO 10
      A(I,NPP)=QJ(II)
      II=II+1
      GO TO 6
   10 A(I,NPP)=0.
    6 CONTINUE
      DO 11 I=NJ,NP
      DO 22 J=1,NP
   22 A(I,J)=0.
      II=I-NJ1
      NNJ=LP(8,II)
      DO 12 J=1,NNJ
      IJ=LP(J,II)
      IIJ=ABS(IJ)
      IF(IJ .LT. 0) GO TO 13
```

```
      A(I,IIJ)=KP(IIJ)
      GO TO 12
   13 A(I,IIJ)=-KP(IIJ)
   12 CONTINUE
   11 A(I,NPP)=0.
      V(1)=4.
C  SYSTEM SUBROUTINE FROM UNIVAC MATH-PACK TO
C  SOLVE LINEAR SYSTEM OF EQ.
     CALL GJR(A,51,50,NP,NPP,$98,JC,V)
      IF (NCT .GT. 0) SUM=0.
      DO 51 I=1,NP
      BB=A(I,NPP)
      IF(NCT) 60,60,61
   60 QM=BB
      GO TO 62
   61 QM=.5*(Q(I)+BB)
      SUM=SUM+ABS(Q(I)-BB)
   62 Q(I)=QM
      DELQ=QM*DELQ1
      QM=ABS(QM)
      V1=(QM-DELQ)/AR(I)
      IF(V1 .LT. .001) V1=.002
      V2=(QM+DELQ)/AR(I)
      VE=QM/AR(I)
      RE1=V1*D(I)/VIS
      RE2=V2*D(I)/VIS
      IF(RE2 .GT. 2.1E3) GO TO 53
      F1=64./RE1
      F2=64./RE2
      EXPP(I)=1.
      KP(I)=64.4*VIS*ARL(I)/D(I)
      GO TO 51
   53 MM=0
      F=1./(1.14-2.*ALOG10(E(I)))**2
      PAR=VE*SQRT(.125*F)*D(I)*E(I)/VIS
      IF(PAR .GT. 65.) GO TO 54
      RE=RE1
   57 MCT=0
   52 FS=SQRT(F)
      FZ=.5/(F*FS)
      ARG=E(I)+9.35/(RE*FS)
      FF=1./FS-1.14+2.*ALOG10(ARG)
      DF=FZ+ELOG*FZ/(ARG*RE)
      DIF=FF/DF
      F=F+DIF
      MCT=MCT+1
      IF(ABS(DIF) .GT. .00001 .AND. MCT .LT. 15) GO TO 52
      IF(MM .EQ. 1) GO TO 55
      MM=1
      RE=RE2
      F1=F
      GO TO 57
   55 F2=F
      BE=(ALOG(F1)-ALOG(F2))/(ALOG(QM+DELQ)-ALOG(QM-DELQ))
      AE=F1*(QM-DELQ)**BE
      EP=1.-BE
```

```
      EXPP(I)=EP+1.
      KP(I)=AE*ARL(I)*QM**EP
      GO TO 51
   54 KP(I)=F*ARL(I)*QM**2
      EXPP(I)=2.
   51 CONTINUE
   17 NCT=NCT+1
C  THE NEXT FIVE CARDS CAN BE REMOVED
      WRITE(6,157) NCT,SUM,(Q( I),I=1,NP)
  157 FORMAT('NCT=',I3,' SUM=',E10.3,/,(1H ,13F10.3))
      WRITE(6,344) (EXPP(I),I=1,NP)
  344 FORMAT(1H  ,13F10.3)
      WRITE(6,344) (KP(I),I=1,NP)
      IF(SUM .GT. ERR .AND. NCT .LT. MAX) GO TO 20
      IF(NCT .EQ. MAX) WRITE(6,108) NCT,SUM
  108 FORMAT('DID NOT CONVERGE IN',I5,' ITERATIONS -
     $SUM OF DIFFERENCES=',E10.4)
      IF(NUNIT .LT. 2) GO TO 63
      WRITE(6,127) (Q(I),I=1,NP)
  127 FORMAT('O FLOWRATES IN PIPES IN CMS',/,(1H  , 3F10.4))
      DO 64 I=1,NP
   64 KP(I)=KP(I)*ABS(Q(I))
      WRITE(6,139) (KP(I),I=1,NP)
      GO TO 30
   63 WRITE(6,107) (Q(I),I=1,NP)
  107 FORMAT('O FLOW RATES IN PIPES IN CFS',/,(1H  , 13F10.3))
      DO 21 I=1,NP
      KP(I)=KP(I)*ABS(Q(I))
   21 Q(I)=449.*Q(I)
      WRITE(6,138) (KP(I),I=1,NP)
  138 FORMAT(' HEAD LOSSES IN PIPES',/,(1H  ,13F10.3))
      WRITE(6,105) (Q(I),I=1,NP)
  105 FORMAT(' FLOW RATES (GPM)',/,(1H  ,13F10.1))
      GO TO 30
   98 WRITE(6,106) JC(1),V
  106 FORMAT(' OVERFLOW OCCURRED -- CHECK
     $SPECIFICATIONS FOR REDUCDANT EQ. RESULTING
     $IN SINGULAR MATRIX',I5,2F8.2)
      GO TO 30
   99 STOP
      END
```

INPUT DATA FOR PREVIOUS 2 LOOP NETWORK PROBLEM

```
   7     6     2    10     0    .001   .00001217    .1
   8.      12.        10.       12.         10.        6.      8.
1106.     751.      1000.      500.        1200.     600.     800.
   .0102    .0102    .0102      .0102       .0102    .0102    .0102
   1     2     1     4
2000.
   0     3    -1    -2     5
   1     3     2    -3    -7
-1500.                            Pipe Nos. at junctions for
   1     2     3    -4            defining continuity equation:
-1000.
   1     2    -5    -6
-1500.
   1     2     6     7
2000.
   4     1    -2    -3    -4      Pipes in loops for defining
   4     5    -6     7     2      energy equation
```

While a short program such as the one above can effectively be used for pipe network analyses, more widely used programs such as those whose input are described in Appendix C include additional features which simplify the input data, and take less computer time for large networks. Features these computer programs may include are: (1) internal checking of errors in input data, (2) internal loop finding algorithms, (3) permitting reservoirs, pumps and pressure reducing valves, (4) banding of the coefficint matrix to save computer storage required and greatly reducing amount of computation required for each iterative solution.

Example Problems Using Linear Theory for Solution

For each of the pipe networks shown below find the flow rate and head loss in each pipe by using the liner theory method to solve the system of equations which considers the flow rates as the unknowns.

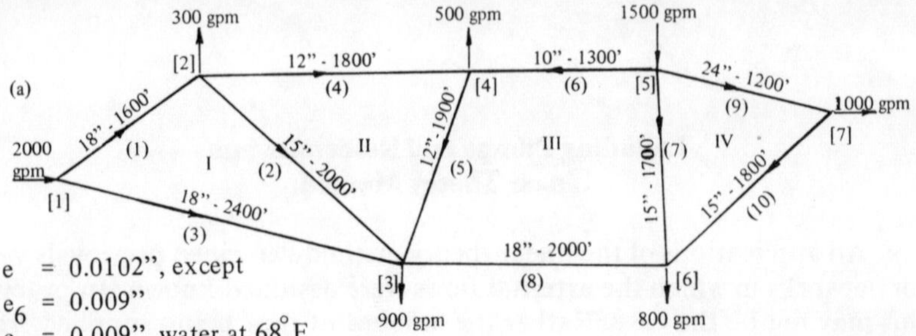

$e = 0.0102"$, except

$e_6 = 0.009"$

$e_8 = 0.009"$, water at 68° F

Answer

Pipe No.	1	2	3	4	5	6	7	8	9	10
Flow rate, Q, cfs	2.19	0.91	2.26	0.61	-0.27	0.23	0.65	-0.90	2.47	0.24
Flow rate, Q (gpm)	984	409	1016	275	-122	102	291	-402	1107	107
Head loss (ft)	0.500	0.297	0.797	0.389	0.092	0.110	0.134	0.116	0.111	0.023

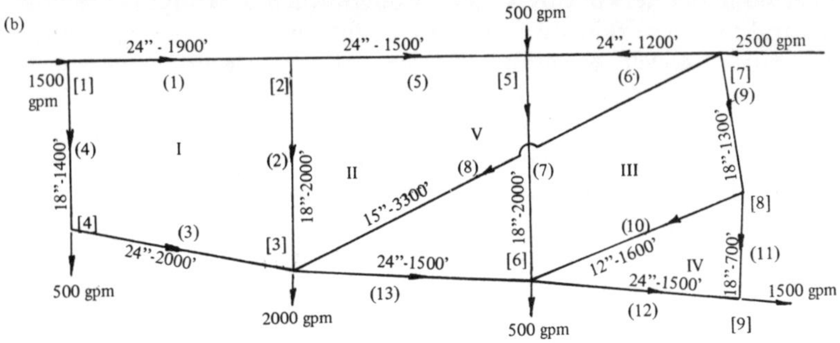

all $e = 0.015"$ except

$e_{10} = 0.01$

$e_{11} = 0.009$

Water at 68° F

Answer

Pipe No.	1	2	3	4	5	6	7
Q (cfs)	0.95	2.03	1.28	2.39	-1.08	2.05	2.09
Q (gpm)	427	912	573	1073	-485	922	937
h_f (ft)	0.031	0.572	0.056	0.546	0.031	0.082	0.602

Pipe No.	8	9	10	11	12	13
Q (cfs)	1.06	2.46	0.39	2.06	1.28	-0.09
Q (gpm)	476	1102	176	927	574	-40
h_f (ft)	0.685	0.535	0.150	0.192	0.042	0.000

Including Pumps and Reservoirs into Linear Theory Method

All applications of the linear theory method described previously were for networks in which the external flows were assumed known. In practice this may not be the case. Rather the amount of flow being supplied from different reservoirs or pumps will depend upon the heads and flows throughout the network. Consequently the utility of the linear theory method can be enhanced by extending it to handle supply sources from reservoirs or pumps, and allow booster pumps to exist within pipelines.

Each pump (not a booster pump) and each reservoir from which flow enters or leaves the system introduces an additional unknown that must be solved for in solving the network. Since pumps and reservoirs must be connected to the network by a pipe through which they supply the flow, it is natural to let the flow rates in these connective pipes be the additional unknowns. However, elevations of reservoirs, and elevations of reservoirs from which pumps obtain the water plus pump characteristics (i.e. h_p versus Q_p) are known for pumps. Therefore, equations are needed which relate these knowns to the connective pipe flow rates. Also for the networks described previously one of the junction continuity equations was redundant, being dependent upon the remaining J-1 junction continuity equations. With pumps and reservoirs supplying the water this is generally no longer the case. (If a junction is assumed at the reservoir or pump a redundant junction continuity equation still occurs. However, reservoirs or pumps will not be considered junctions.) Consequently, if one resevoir and one pump supply the flow to the network, as illustrated in the sketch below, such that flows in two connective pipes become additional unknowns, one additional equation is necessary beyond the J available continuity equations and the L available energy equations. A convenient means for obtaining this additional equation is by defining a so called "pseudo loop," which connects the two reservoirs (in this instant the reservoir supplying flow and the reservoir from which the pump obtains flow) by a "no flow" pipe as illustrated below. Note that two pumps and/or reservoirs must be present or the network reduces to one for which all external flows (i.e. supplies and consumptions) are known. Consequently such pseudo loops can always be defined, because at least two reservoirs and/or pumps must exist in a network if all external flow rates are not known. The additional needed equation (or equations if more than two pumps and/or reservoirs are present), comes from the energy equation around this pseudo loop. Around pseudo loops the sum of head loss is not equal to zero but must account for the difference in reservoir elevations and head produced by the pump or pumps. Around the pseudo loop in the diagram below the energy equation is,

$$H_2 - K_5 Q_5^{n_5} - K_4 Q_4^{n_4} + K_3 Q_3^{n_3} + K_2 Q_2^{n_2}$$
$$+ K_1 Q_1^{n_1} - h_p = H_1$$

in which h_p is the head produced by the pump and other symbols are as previously defined. Writing this equation so that head losses in the direction of flow are positive, to be consistent with previous equations, gives.

$$- K_1 Q_1^{n_1} - K_2 Q_2^{n_2} - K_3 Q_3^{n_3} + K_4 Q_4^{n_4} + K_5 Q_5^{n_5}$$
$$+ h_p = H_2 - H_1$$

or in general

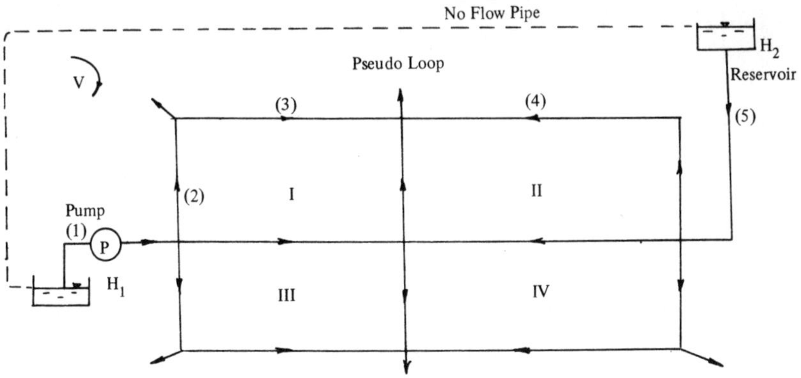

$$\Sigma K_i\, Q_i^{n_i} \;\pm \Sigma\, h_p \;=\; \Delta H \qquad \text{(5-3)}$$

represents the energy equation around a pseudo loop containing two reservoirs and/or source pumps.

A number of alternative methods might be used to quantify the head h_p produced by the pump. The method used herein will approximate h_p over its working range by a quadratic equation of the form

$$h_p \;=\; AQ^2 + BQ + H_o \qquad \text{(5-4)}$$

in which A, B, and H_0 are constants for a given pump and might be determined by fitting Eq. 5-4 to three points taken from a pump characteristic curve.

When Eq. 5-4 is substituted into Eq. 5-3 a nonlinear equation results which contains only flow rates in pipes of the network (including connective pipes) as the unknowns. In this form the linear theory method does not give rapid convergence as it does when pumps and/or reservoirs are not present. The system of equations will therefore be modified to allow the linear theory method to converge rapidly. To help appreciate the motivation for this modification, reasons why the presence of pumps can greatly increase the number of required iterations in the linear theory method will be investigated.

Typical centrifugal pump characteristic curves are shown below. If the exponental formula is fit through the two points shown, the following equation results.

$$h_p \;=\; 15.84\, Q^{-0.503}$$

Note the negative exponent -0.503 is vastly different from the exponents in the expontial formula for pipes which are typically slightly less than two.

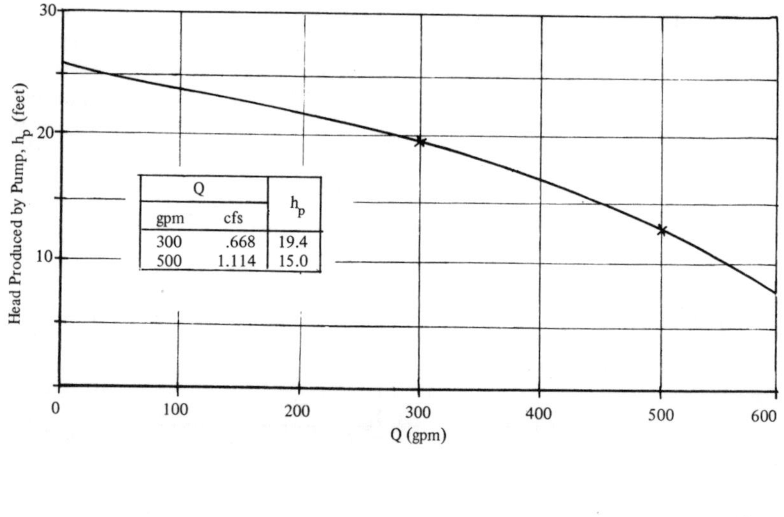

The head produced by this pump decreases nearly proportional to the reciprocal of the square root of the flow rate whereas the head loss in a typical pipe increases approximately proportional to the square of the flow rate. Clearly these relationships are quite dissimilar. A consequence is that when typical pump characteristic curves are used to define h_p in Eq. 5-3 and this equation becomes one of the simultaneous equations solved by the linear theory method, converence may become very slow if at all.

To improve this situation a transformation of variables is needed so that the unknown which replaces h_p in Eq. 5-3 has an exponent value close to the rest of the n's. Such a transformation is

$$G = Q + \frac{B}{2A} \quad \text{.......................(5-5)}$$

in which G is the new variable and A and B are the constants in Eq. 5-4. To demonstrate the appropriateness of Eq. 5-5, solve it for Q and substitute into Eq. 5-4. After some simplification

$$h_p = AG^2 + h_o \quad \text{.......................(5-6)}$$

in which

$$h_o = H_o - \frac{B^2}{4A} \quad \text{.......................(5-7)}$$

Obviously the exponent of G (i.e. 2) is close to the value of a typical n in the exponential formula. Substituting Eq. 5-6 for h_p in Eq. 5-3 and moving the h_o's to the other side of the equal sign gives

$$\text{.............} \quad \Sigma K_i Q_i^{n_i} \mp \Sigma AG^2 = \Delta H + \Sigma h_o \quad \text{..............(5-8)}$$

Each term in Eq. 5-8 is similar to typical terms in the energy equation written around real loops. The only problem now is that for each pump an additional unknown G has been introduced. However, the transformation equation, Eq. 5-5, is a linear equation which relates G to the flow rate in the line containing the pump to the network. By adding these additional equations to the system as many equations are produced as unknowns exist and a solution can be obtained. This new system of equations does converge rapidly to the solution by the linear theory method.

In summary then, if pumps or reservoirs exist in a pipe network, a solution by the linear theory method is accomplished as follows:

1. J linear junction continuity equations are written.
2. L nonlinear energy equations are written around real natural loops of the system.
3. Additional pseudo loops are defined which connect supply reservoirs or reservoirs from which pumps obtain their supply by no flow pipes. Energy equations are written around these pseudo loops. These energy equations contain new unknown G_i for each pump in the net-

work. The number of these pseudo loops must equal the difference between the number of unknown flow rates, i.e. N and (J + L).

4. As many additional linear equations of the form G - Q = B/2A (Eq. 5-5) are written as pumps exist.
5. The nonlinear energy equations are linearized by defining coefficients K' of the Q's which are obtained by $K' = KQ(m)^{n-1}$ and coefficients K'_G for the G unknowns are obtained by $K'_G = AG$.
6. The resulting system is solved iteratively, adjusting the coefficients as described earlier to reflect the average of the flow from the past two solutions until convergence occurs.

Should any of the details involved in these steps be vague, following them through for a simple example will be helpful. Consider the seven-pipe, one loop network supplied by three identical pumps shown below. Each pump supply head according to the equation

$$h_p = -10.328\,Q_p^2 + 2.823\,Q_p + 22.289$$

Since there are seven pipes in this network there will be seven unknown flow rates, plus three additional unknowns, i.e. the G's of Eq. 5-5 for the three pumps which supply flow. Consequently a total of 10 simultaneous equations are needed. Four of these equations are the junction continuity equations; three are from Eq. 5-5 relating the three G's to Q_1, Q_5, and Q_7; and consequently three energy equations are needed, one from the real loop and two from pseudo loops connecting pump reservoirs with no flow pipes. With the K's in the exponential formula approximately computed by the Hazen-Williams equation, these equations are:

$$\text{Continuity Equations}\quad \begin{cases} -Q_1 + Q_2 + Q_6 = 0 \\ -Q_2 - Q_3 = -2.0 \\ Q_3 - Q_4 - Q_5 = 0 \\ Q_4 - Q_6 - Q_7 = 0 \end{cases}$$

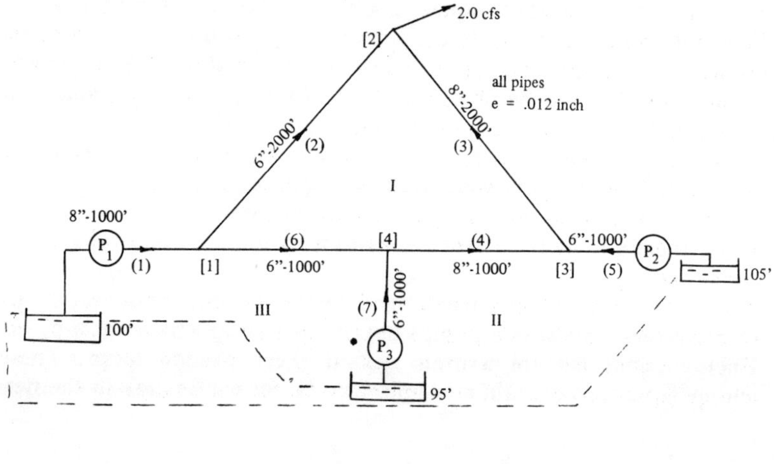

Real Loop
$$\begin{cases} 55.7\,Q_2^{1.85} - 13.7\,Q_3^{1.85} - 6.86\,Q_4^{1.85} \\ -27.8\,Q_6^{1.85} = 0 \end{cases}$$

Pseudo Loop
$$\begin{cases} 6.86\,Q_1^{1.85} + 6.86\,Q_4^{1.85} - 27.8\,Q_5^{1.85} \\ + 27.8\,Q_6^{1.85} + 10.33\,G_1 - 10.33\,G_2 = -5 \\ 6.86\,Q_1^{1.85} + 27.8\,Q_6^{1.85} - 27.8\,Q_7^{1.85} \\ + 10.33\,G_1 - 10.33\,G_3 = 5 \end{cases}$$

Transformation Eq. 5-5
$$\begin{cases} -Q_1 + G_1 = -0.137 \\ -Q_5 + G_2 = -0.137 \\ -Q_7 + G_3 = -0.137 \end{cases}$$

In applying the linear theory method, the three nonlinear energy equations are linearized as described previously, and the resulting linear system solved. After three such iterative solutions of the linearized system the following solution results:

Pipe No.	1	2	3	4	5	6	7
Flow rate (cfs)	0.841	0.666	1.334	0.573	0.761	0.175	0.398
Head loss	3.15	17.63	15.46	1.50	11.42	0.674	3.24

Junction No.	1	2	3	4
Head (ft)	114.2	96.6	112.0	113.5
Pressure (psi)	49.5	41.9	48.5	49.2

Pump	Head (ft)
1	17.36
2	18.46
3	21.78

Including Pressure Reduction Valves Into Linear Theory Method

A pressure reducing valve (denoted PRV) is designed to maintain a constant pressure downstream from it regardless of how large the upstream pressure is. The exceptions to this occurrence are: (1) If the upstream pressure becomes less than the valve setting, and (2) if the downstream pressure exceeds the pressure setting of the valve so that if the PRV were not present the flow would be in the opposite direction to the downstream flow direction of the valve. If the first condition occurs, the valve has no effect on flow conditions. The PRV acts as a check valve, preventing reverse flow if the second condition occurs. By preventing reverse flow, the PRV allows the pressure immediately downstream from the valve to exceed its pressure setting. Thus, PRV's are used to reduce pressures in portions of a pipe distribution system if the pressures would otherwise be excessive, and they may also be used to control from which sources of supply the flow comes under various demand levels. In the latter applications the PRV acts as a check valve until the pressure is reduced to critical levels by large demands at which time additional sources of supply are drawn upon.

The analysis of a pipe network containing one or more PRV's must be capable of determining which of these conditions exist. Methods for accomplishing this which are consistent with the linear theory method are discussed in this section. Problems 11 through 15 in the next section provide additional examples, beyond those given in this section, of small networks containing PRV's.

Example No. 1 Methods for including PRV's in network analysis by the linear theory method will be illustrated by several examples. First consider the 7-pipe network shown below in which the a PRV exists in pipe

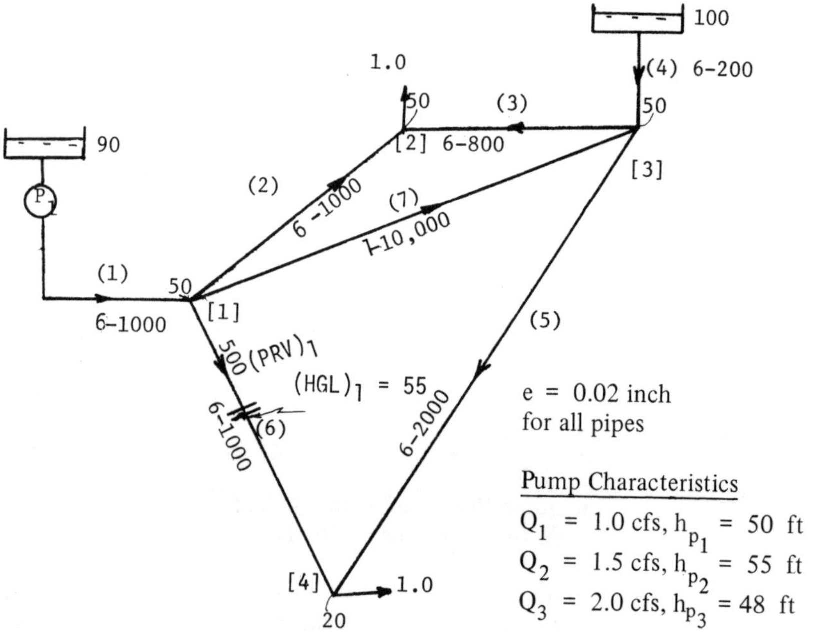

6, 500 feet downstream from the beginning of this pipe. The system of Q-equations for this network consists of four junction continuity equations and three energy equations obtained from loops. The junction continuity equations are identical to those that would be written if the PRV were not present. In forming the loops for the energy equations, however, the pipe containing the PRV (pipe 6) is disconnected from its upstream junction and the PRV is replaced by a reservoir. (This assumes the PRV reduces the pressure to a constant value. The PRV acting as a check valve is discussed later.) Of the three loop equations, one is obtained by summing the head losses around the real loop formed by pipes 2, 3, and 7. The second equation comes from the pseudo loop which connects the upper reservoir to the reservoir from which the pump obtains water (i.e., the source pump) by the pipes 4, 7, and 1. The third loop equation connects the artificial reservoir created by the PRV to another reservoir by a series of pipes such as 6, 5, and 4. Note that with pipe 6 disconnected from junction 1, only one real loop is availabe whereas two independent real loops existed before this. The real loop which is lost through the disconnection is compensated for by the additional pseudo loop from the artificial reservoir created by the PRV. Therefore, the number of equations which are available always equals the number of unknown flow rates Q. Using this scheme, the eight equations (eight are used instead of seven because one equation is added by the pump transformation as described in the previous section) needed for a solution by the linear theory method are:

$$\left.\begin{aligned} -Q_1 + Q_2 + Q_6 + Q_7 &= 0.0 \\ -Q_2 - Q_3 &= -1.0 \\ Q_3 - Q_4 + Q_5 - Q_7 &= 0.0 \\ -Q_5 - Q_6 &= -1.0 \end{aligned}\right\} \text{Junction Continuity Equations}$$

$$-K_2 Q_2^{n_2} + K_3 Q_3^{n_3} + K_7 Q_7^{n_7} = 0.0 \qquad \text{Real Loop}$$

$$K_4 Q_4^{n_4} - K_7 Q_7^{n_7} - K_1 Q_1^{n_1} + A_1 G_1^{2} = 100 - 90 - h_{01} \qquad \text{Pseudo Loop}$$

$$G_1 - Q_1 = (B/2A)_1 \qquad \text{Pump Transformation}$$

$$K_6 Q_6^{n_6} - K_5 Q_5^{n_5} - K_4 Q_4^{n_4} = 55 - 100 \qquad \text{Pseudo Loop from PRV}$$

in which K_6 is determined only for the portion of pipe 6 downstream from the PRV.

Upon solving this system of equations by the linear theory method, using the procedure described previously, the following solution results:

Pipe No.	1	2	3	4	5	6	7
Flow rate (cfs)	1.03	0.98	0.017	0.97	0.96	0.041	0.003
Head loss (ft)	23.35	21.43	0.009	4.21	40.92	0.05	23.68

The hydraulic grade line elevations at the junctions are:

Junction No.	1	2	3	4
Elevation of HGL (ft)	117.21	95.79	95.79	57.98

The pressure upstream from the PRV equals 117.19 ft. and downstream equals 55 ft. Consequently, the assumption used in writing the final loop equation above is correct. Had the solution given a negative flow rate in pipe 6, this assumption would be incorrect, since the PRV would then have acted as a check valve and allowed the elevation of the HGL downstream from the PRV to rise above 55 ft. Should this have occurred, then the flow rate in pipe 6 would no longer be unknown, but equal to zero. Under these conditions caused by the PRV shutting flow off, the final equation above would be replaced by:

$$-(HGL)_1 - K_5 Q_5^{n_5} - K_4 Q_4^{n_4} = -100$$

From the above example, it should be evident how networks containing PRV's might be analyzed by the linear theory method. The assumption is initially made that the pressure (or the elevation of the HGL) immediately downstream from the PRV is constant and equal to the valve setting. The junction continuity equations are written as if no PRV's existed. To write the loop equations, pipes containing PRV's are disconnected from their upstream junctions and the PRV's are replaced by reservoirs. The real and pseudo loops are established for this modified network. After each iteration a check of the flow Q in each pipe containing a PRV is made. Should any of these flows be negative, then the pseudo loop equation which includes terms for that pipe is altered, replacing that Q as an unknown with the HGL immediately downstream from that PRV.

Example No. 2 As an example of a network in which a PRV acts as a chech valve, consider the 14-pipe network in the sketch below. In this network, pipe 5 contains a PRV with a pressure setting which results in

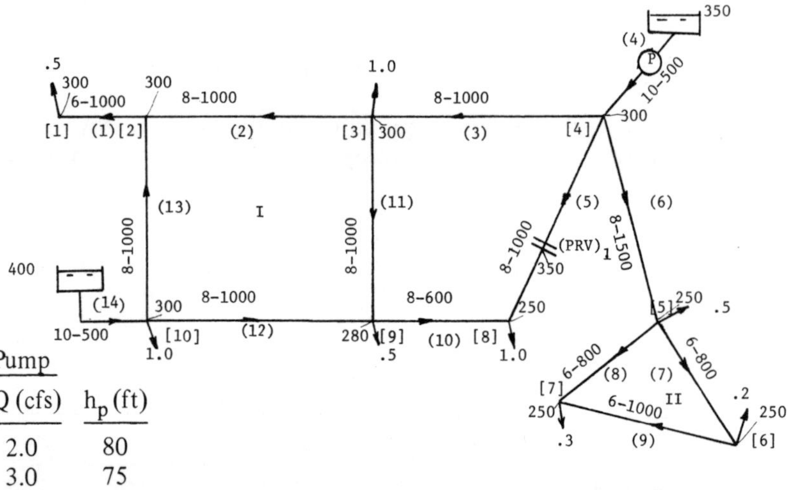

Pump

Q (cfs)	h_p (ft)
2.0	80
3.0	75
4.0	68

the elevation of the HGL downstream from the valve equaling 350 ft. For this network 10 linear junction continuity equations are available, and two energy equations around the real loops formed by pipes 2, 11, 12, and 13, and 8, 9, and 7, respectively. One additional energy equation is available from a pseudo loop connecting the reservoir to the source pump through a series of pipes such as 14, 13, 2, 3, and 4. The tenth and final equation is obtained by connecting the artificial reservoir created by the PRV to the real reservoir or source pump by a series of pipes such as 5, 10, 12, and 14. This final equation is,

$$K_5 Q_5^{n_5} - K_{10} Q_{10}^{n_{10}} - K_{12} Q_{12}^{n_{12}} - K_{14} Q_{14}^{n_{14}} = (HGL)_1 - 400$$

Initially, $(HGL)_1$ might be assumed known and equal to 350 ft. If this is done, however, the linear theory method immediately produces a negative flow rate, Q_5, from the first iteration. Consequently $Q_5 = 0$ and the above equation should be rewritten with $(HGL)_1$ unknown as,

$$-(HGL)_1 - K_{10} Q_{10}^{n_{10}} - K_{12} Q_{12}^{n_{12}} - K_{14} Q_{14}^{n_{14}} = -400$$

The solution obtained using this equation as the fourteenth equation of the system of equations is

Pipe No.	1	2	3	4	5	6	7
Flow rate (cfs)	0.50	0.37	2.16	3.16	0.0	1.00	0.25
Head loss (ft)	2.44	0.63	19.27	6.40	0.0	6.40	1.01

Pipe No.	8	9	10	11	12	13	14
Flow rate (cfs)	0.25	0.05	1.00	0.79	0.71	0.13	1.84
Head loss (ft)	1.06	0.06	2.56	2.73	2.22	0.09	2.21

The elevations of the HGL upstream and downstream from the PRV are 417.60 ft and $(HGL)_1 = 393.04$ ft, respectively.

While it is possible to obtain a solution to a network such as this 14-pipe network in a single analysis, it might be analyzed piecemeal more efficiently. Since all the demands at junctions 5, 6, and 7 must come through pipe 6, its flow rate can be determined readily as the sum of these demands or $Q_6 = 1.0$ cfs. Now the subnetwork consisting of pipes 7, 8, and 9 can be disconnected from the remainder of the network and each subnetwork analyzed separately. When large distribution systems are involved, considerable reduction in storage requirements as well as computational efficiency can result by disconnecting the network into subnetworks where possible.

Often PRV's are installed in a series of pipes to prevent excessive pressures in a lower elevation portion of a city. Such series of PRV's may cause pressures in a portion of the network to be completely independent of the remainder of the network. When this occurs, it is best to form subnetworks and analyze each separately starting with the lower elevation subnetworks. The solution from the isolated lower elevation subnetwork is used to determine the demands at the various junctions of the higher elevation subnetwork.

The small 10-pipe network containing three PRV's in pipes 4, 5, and 7, respectively, illustrates such a situation. In this network, the three

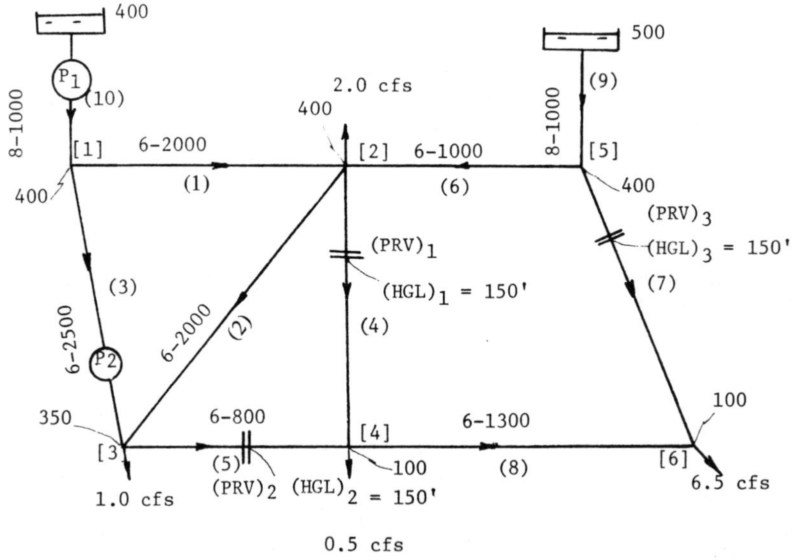

PRV's cause the pressures at junctions 4 and 6 to be independent of the pressures throughout the remainder of the network. The best method of analysis, therefore, consists of first solving the subnetwork consisting of pipes 5, 4, 7, and 8 downstream from the PRV's. In this subnetwork of four pipes, the three PRV's are handled as three constant head resevoirs. The values for Q_4, Q_5, and Q_7 from the solution of this 4-pipe subnetwork are next added to the other demands to obtain the demands at junctions 3, 2, and 5, respectively, in setting up the remainder of the network for a solution.

While it is generally not difficult to determine whether PRV's isolate a portion of a network from the rest by visual examination of its sketch, a simple test is needed for determining this for use in a computer solution. Such a test can be based on the fact that no series of connected pipes exist between any of the artificial reservoirs created by the PRV's and the other supply reservoirs. That no such connection exists in this 10-pipe network can be seen by resketching the network as shown below. Thus, if pseudo loops between artificial reservoirs and source reservoirs cannot be found in a computer program with its own internal loop findings algorithms, then the PRV's isolate a subnetwork from the remainder of the network.

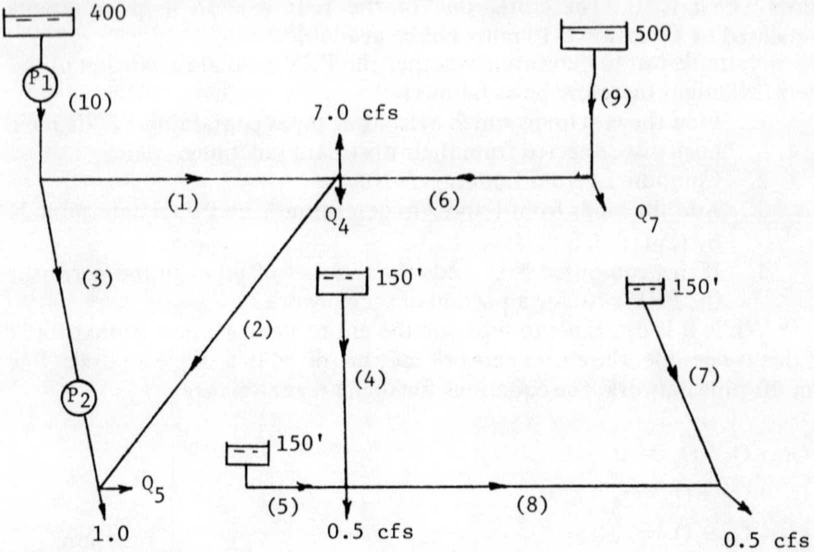

An alternate to this test is also available. This alternate is derived from the fact that the number of pipes in a network supplied by reservoirs and/or source pumps equals the number of junctions plus the number of independent loop equations (the natural real loops plus the pseudo loops), or

$$N = J + L \qquad (5\text{-}9)$$

The number of pseudo loop equations generally available equals the sum of reservoirs and source pumps minus one or

$$L_s = N_r + N_{p_s} - 1 \qquad (5\text{-}10)$$

If the number of pseudo loops L_S plus the number of real loops plus the number of junctions exceeds the number of pipes, then one of these equations is not available. In other words, this sum of equations exceeds the number of unknowns; an impossibility if the equations are independent. Therefore, one of these equations is not available, i.e., one or more of the equations from the pseudo loops is not available. The equation is not availble since no series of connected pipes exists between PRV's and other reservoirs.

To illustrate the above, if the three PRV's in the 10-pipe network above are considered reservoirs, then the number of reservoirs $N_r = 4$. Since one source pump exists ($N_{ps} = 1$), Equation 5-10 would indicate $L_s = 4$. Since one real loop exists (pipes 1, 2, and 3) $L = 5$. Since $J = 6$, Equation 5-9 gives $N = 6+5 = 11$ which is larger than the number of pipes which is 10. Therefore, one of the four pseudo loop equations indicated by Equation 5-10 must not be available.

A simple test to determine whether the PRV's isolate a portion of the network might therefore be as follows:

1. Find the real loops which exist after pipes containing PRV's have been disconnected from their upstream junctions.
2. Compute L_s from Equation 5-10.
3. Add the loops from 1 and 2 to determine L and then determine N by Equation 5-9.
4. If this computed N exceeds the number of pipes in the network, the PRV's isolate a portion of the network.

While it is desirable to separate the entire network into subnetworks if this is possible, the entire network may be solved in a single analysis. For the 10-pipe network, the equations for such an analysis are:

$$\left.\begin{aligned}
&Q_1 + Q_3 - Q_{10} = 0\\
&-Q_1 + Q_2 + Q_4 - Q_6 = -2.0\\
&-Q_2 - Q_3 + Q_5 = -1.0\\
&-Q_4 - Q_5 + Q_8 = -0.5\\
&Q_6 + Q_7 - Q_9 = 0.0\\
&-Q_7 - Q_8 = -0.5
\end{aligned}\right\}\ \text{Junction Continuity}$$

$$K_3 Q_3^{n_3} - K_2 Q_2^{n_2} - K_1 Q_1^{n_1} = 0 \qquad \text{Real Loop}$$

$$\left.\begin{aligned}
&K_{10} Q_{10}^{n_{10}} + K_1 Q_1^{n_1} - K_6 Q_6^{n_6} - K_9 Q_9^{n_9} - h_{p_1} = 400-500\\
&K_4 Q_4^{n_4} + K_8 Q_8^{n_7} - K_7 Q_7^{n_7} = (HGL)_1 - (HGL)_3 = 0\\
&K_5 Q_5^{n_5} + K_8 Q_8^{n_8} - K_7 Q_7^{n_7} = (HGL)_2 - (HGL)_3 = 0
\end{aligned}\right\}\ \text{Pseudo Loop}$$

Note that none of the pseudo loops connect the artificial reservoirs to real reservoirs. Rather, the last two of the above equations are obtained from pseudo loops, one PRV to another PRV. The solution to the above system of equation (when K_i and n_i are defined from the data given on the sketch) provides the following.

Pipe No.	1	2	3	4	5
Flow rate (cfs)	0.976	0.172	1.266	0.236	0.438
Head loss (ft)	35.92	1.28	74.79	2.32	3.02

Pipe No.	6	7	8	9	10
Flow rate (cfs) Head loss (ft)	1.432 38.15	0.326 3.86	0.174 0.85	1.758 12.83	2.242 20.70

PRV No.	1	2	3
Upstream elevation of HGL (ft) Downstream elevation of HGL (ft)	448.21 150.00	446.23 150.00	485.67 150.00

Example Problems Which Include Pumps and Reservoirs

1. Water supply comes from one pump and two reservoirs as shown in the sketch. How many pseudo loops need to be established? Write the system of equations whose solution provides the flow rate in each pipe using symbols K and n in the energy equations. The pump characteristic curve is given by: $h_p = -10.33\,Q_p^2 + 2.823\,Q_p + 22.29$.

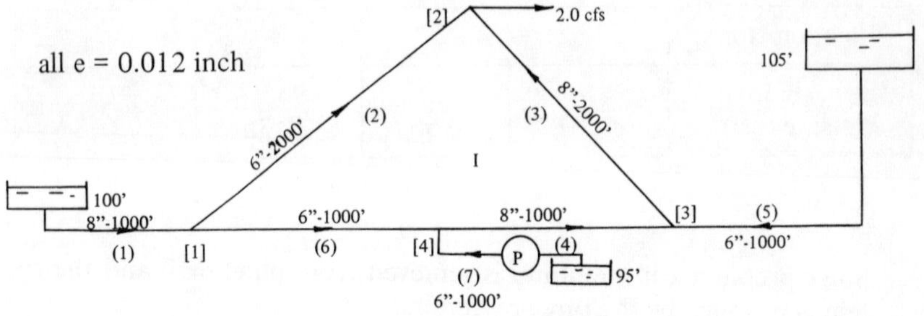

Solution:

Two pseudo loops are required. A possibility is one pseudo loop connecting the reservoirs supplying pipes 1 and 5 through pipes 1, 6, 4, and 5; and the other connects the pump reservoir and the reservoir supplying pipe 1 through pipes 1, 6, and 7.

Continuity Equations
$$\begin{cases} -Q_1 + Q_2 + Q_6 = 0 \\ -Q_2 - Q_3 = -2.0 \\ Q_3 - Q_4 - Q_5 = 0 \\ Q_4 - Q_6 - Q_7 = 0 \end{cases}$$

Real Loop
$$\left\{ K_2 Q_2^{n_2} - K_3 Q_3^{n_3} - K_4 Q_4^{n_4} - K_6 Q_6^{n_6} = 0 \right.$$

Pseudo Loop

$$\begin{cases} K_1Q_1^{n_1} + K_6Q_6^{n_6} + K_4Q_4^{n_4} - K_5Q_5^{n_5} = 5 \\ K_1Q_1^{n_1} + K_6Q_6^{n_6} - K_7Q_7^{n_7} + 10.33\,G_1 \\ \quad = -5 + 22.36 \end{cases}$$

Transformation Equation

$$\{ -Q_7 + G_1 = -0.137$$

2. Solve the network of problem 1 giving the flow rates in each pipe, the head loss in each pipe, and the head and pressures at each junction, if the elevation of all junctions is at 80 ft.

Solution:

Pipe No.	1	2	3	4	5	6	7
Flow rate (cfs)	0.533	0.662	1.338	0.699	0.639	-0.129	0.828
Head loss (ft)	1.306	17.384	15.563	2.201	8.127	0.380	13.474

Junction No.	1	2	3	4
Elev. HGL	98.7	81.3	96.9	99.1
Pressure (psi)	8.10	0.56	7.32	8.28

3. Solve problem 1 if the pump is removed from pipeline 7 and the system is supplied by the three reservoirs.

Solution:

To solve this system there will only be seven unknowns instead of the eight as in problems 1 and 2, since the pump does not introduce an additional unknown G. The four continuity equations and the first two energy equations are identical to those given in the solution to problem 1. The final energy equation is,

$$K_1Q_1^{n_1} + K_6Q_6^{n_6} - K_7Q_7^{n_7} = -5$$

The solution by the linear theory method produces the following after three iterative solutions.

Pipe No.	1	2	3	4	5	6	7
Q (cfs) h_f (ft)	0.979 4.23	0.680 18.32	1.320 15.16	0.527 1.28	0.793 12.39	0.300 1.88	0.227 1.11

4. Solve the single real loop pipe network supplied by the three pumps described just before these example problems by the Hazen-Williams equation assuming all pipes have a coefficient $C_{HW} = 120$.

Solution:

The equations are the same as previously given with the exception that the n's = 1.852, and the K's are constant and given by: $K_1 \doteq 4.85$, $K_2 = 39.3$, $K_3 = 9.69$, $K_4 = 4.85$, $K_5 = 19.7$, $K_6 = 19.7$, $K_7 = 19.7$. The solution is

Pipe No.	1	2	3	4	5	6	7
Q (cfs) h_f (ft)	0.784 3.09	0.681 19.3	1.313 16.1	0.564 1.68	0.751 11.6	0.257 1.59	0.368 3.09

5. Solve problem 3 using the Hazen-Williams formula with all C_{HW} = 120.

Solution:

Pipe No.	1	2	3	4	5	6	7
Q (cfs) h_f (ft)	0.965 4.53	0.684 19.45	1.316 16.13	0.522 1.45	0.795 12.86	0.281 1.87	0.241 1.41

6. Obtain a solution to the 28 pipe network below by the linear theory method through use of a program such as the one whose input is described in Appendix C.

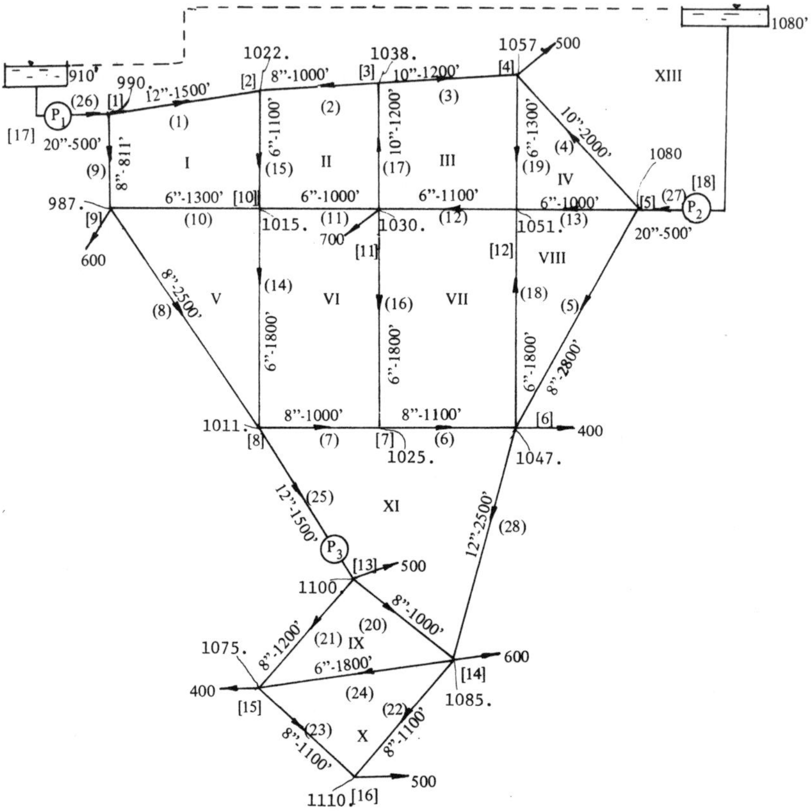

Pipe No.	C_{HW}	Pipe No.	C_{HW}
1	130	15	120
2	130	16	120
3	120	17	130
4	120	18	120
5	120	19	120
6	120	20	130
7	120	21	120
8	120	22	120
9	100	23	120
10	100	24	120
11	100	25	130
12	130	26	110
13	130	27	110
14	120	28	120

Pump Characteristics

$$\text{Pump No. 1 } h_p = -0.417\, Q_p^2 + 5.14\, Q_p + 440.3$$
$$\text{Pump No. 2 } h_p = -0.378\, Q_p^2 - 5.09\, Q_p + 298.2$$
$$\text{Pump No. 3 } h_p = -2.51\, Q_p^2 + 16.7\, Q_p + 155.3$$

Data for program:

```
    O PROBLEM 6, CHAPTER 5
    $SPECIF NPIPE=28,NRES=1,NPUMP=3,NBPUMP=1,NFLOW=1,COEFRO=120.,
     NPLENG=1 $END
    27
    1080.
26 27 25
5.568 455.97 7.0156 455.81 7.795 455. 910.
 2.004 280.5 2.3385 284.25 3.341 277. 1080.
4.009 182. 4.454 180. 4.9 177. 0.
1 12.   1.5   130.  1  2  1022.
2  8. 1. 130.  3  2
3  10.   1.2    3  4  500.  1057
4  10.   2.  5  4
5  8.  2.8    5  6  400.  1047.
6  8.  1.1    7  6
7  8.  1.  8  7
8  8.  2.5    9  8  1011.
9  8.  .811  100.  1  9  600.  987.
10 6.  1.3   100.  9  10 1015.
11 6.  1.  100.  10 11 700.  1030.
12 6.  1.1    130.  12 11
13 6.  1.  130.  5  12 1051.
14 6.  1.8    10 8
15 6.  1.1    2  10
16 6.  1.8    11 7  1025.
17 10.   1.2    130.  11 3  1038.
18 6.  1.8    6  12
19 6.  1.3    4  12
20 8.  1.  130.  13 14 600.  1085.
21 8.  1.2    13 15 400.  1075.
22 8.  1.1    14 16 500.  1110.
23 8.  1.1    15 16
24 6.  1.8    14 15
26 20.   .5  110.  1  990.
27 20.   .5  110.  5  1080.
28 12.   2.5    6  14
END
```

Solution:

One pseudo loop must be defined which connects the reservoirs from which pumps 1 and 2 obtain their supplies, giving 16 junction continuity equations and 12 energy equations. In addition three linear transformation equations are added for the three pumps in the network. The data required by the program in Appendix C follows. The solution is provided below.

Pipe No.	1	2	3	4	5	6	7
Q (cfs) h_f (ft)	2.94 6.41	-1.76 11.85	-0.54 0.62	1.74 9.10	0.88 10.76	-2.55 30.2	-3.35 45.5

Pipe No.	8	9	10	11	12	13	14
Q (cfs) h_f (ft)	2.17 51.1	3.07 43.9	-0.44 7.92	-0.58 9.90	0.64 8.20	0.73 9.38	1.32 59.0

Pipe No.	15	16	17	18	19	20	21
Q (cfs) h_f (ft)	1.18 29.6	0.80 23.4	-2.29 7.86	-0.17 1.37	0.09 0.28	3.27 37.6	2.45 30.7

Pipe No.	22	23	24	25	26	27	28
Q (cfs) h_f (ft)	-0.04 0.01	1.15 6.90	-0.41 6.91	6.84 30.6	6.01 0.91	3.35 0.31	-2.39 8.43

Junction No.	1	2	3	4	5	6	7	8
Elev. HGL (ft)	1365	1359	1347	1348	1357	1346	1316	1270

Junction No.	9	10	11	12	13	14	15	16
Elev. HGL (ft)	1321	1329	1339	1347	1392	1354	1361	1354

7. What are the flow rates and head losses, as well as the elevation of the HGL in the network of problem 6 if pipeline 28 is taken out of operation. If the elevation at junctions are as given below in one of the tables of solutions what is the pressure at each junction. The solution should be obtained by the computer program such as those whose input is described in Appendix C.

Solution:

With pipeline 28 removed a single line connects two separate networks, each of which might be solved as separate problems. Obviously, if the analysis were being done by hand two separate analyses would be used and subsequently tied together. With the computer, however, it is more convenient to consider only one network containing 27 pipes, 16 junctions, and 10 real loops. The additional energy equation comes from defining a pseudo loop between the reservoirs from which pumps 1 and 2 obtain their supply. Pump 3 is a booster pump and therefore no pseudo loop needs to include it, but it adds an additional unknown G_3 giving a total of 30 unknowns and 30 simultaneous equations to solve by the linear theory method. The answers are:

Pipe No.	1	2	3	4	5	6	7
Q (cfs)	2.92	-1.90	-0.19	1.60	1.43	-1.18	-1.90
h_f (ft)	6.35	13.76	0.09	7.83	26.3	7.28	15.89

Pipe No.	8	9	10	11	12	13	14
Q (cfs)	1.60	2.62	-0.31	-0.25	0.43	0.78	0.96
h_f (ft)	28.79	32.88	4.06	2.09	3.95	10.60	32.85

Pipe No.	15	16	17	18	19	20	21
Q (cfs)	1.02	0.72	-2.09	-0.65	0.30	1.83	1.52
h_f (ft)	22.47	19.04	6.63	15.71	2.77	12.73	12.57

Pipe No.	22	23	24	25	26	27
Q (cfs)	0.54	0.57	-0.06	4.45	5.55	3.81
h_f (ft)	1.72	1.89	0.16	13.84	0.78	0.39

Junction No.	1	2	3	4	5	6	7	8
Elev. (ft)	990	1022	1038	1057	1080	1047	1025	1011
HGL (ft)	1365	1359	1345	1345	1353	1327	1319	1303
Pressure (psi)	163	146	133	125	118	121	128	127

Junction No.	9	10	11	12	13	14	15	16
Elev. (ft)	987	1015	1030	1051	1100	1085	1075	1110
HGL (ft)	1332	1336	1338	1342	1470	1457	1458	1455
Pressure (psi)	150	139	134	126	160	161	166	150

8. Solve the 51 pipe networks shown below using a computer program. All pipes have a wall roughness e = 0.012 inch. Three points along the pump characteristics are $Q_{p1} = 0$, $h_{p1} = 196'$; $Q_{p2} = 20.43$ cfs, $h_{p2} = 151.37$, $Q_3 = 25.0$ cfs, $h_{p3} = 44.0'$.

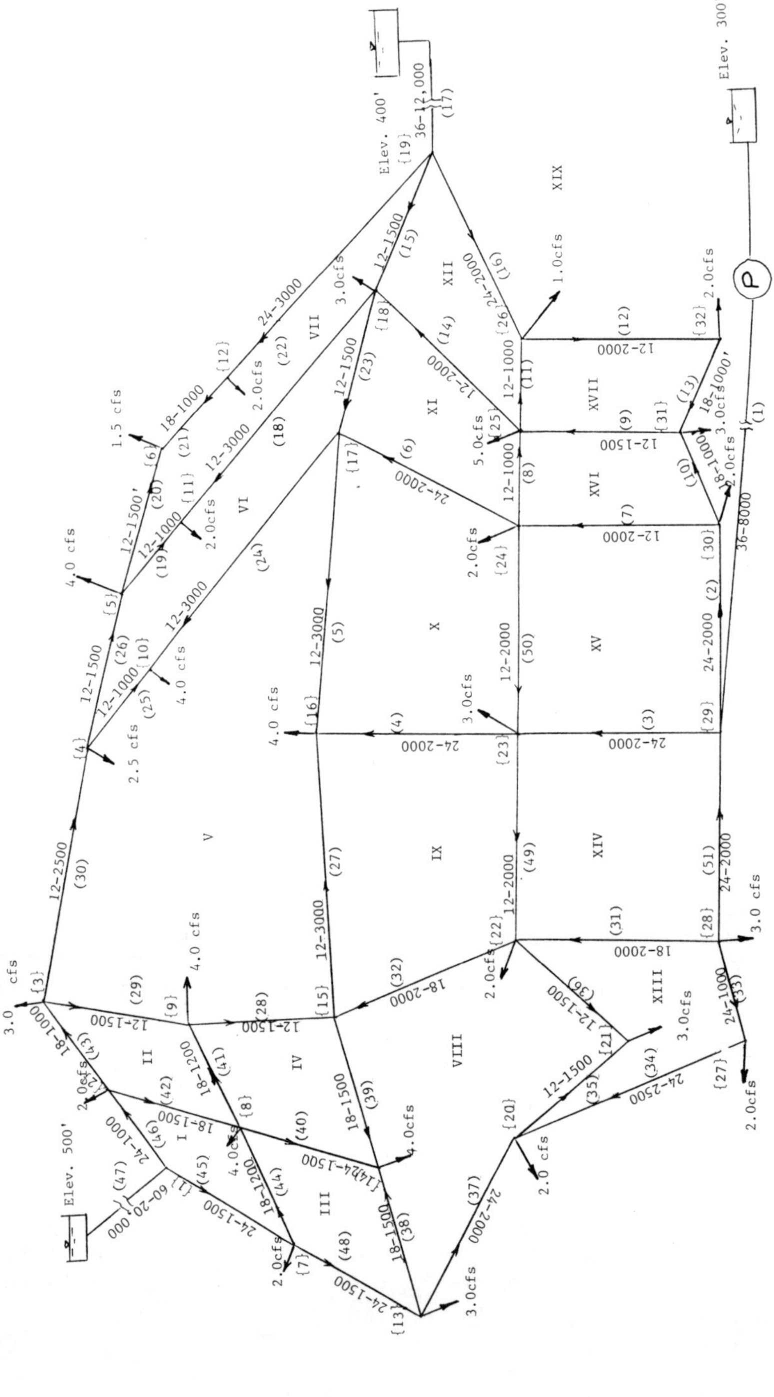
Elev. 400'
36-12,000
(17)
{19}
Elev. 300'
P
36-8000
(1)
Elev. 500'
60-20 000
(47)
24-3000
18-1000
12-1500
12-2500
18-1000
24-1000
24-1500
18-1500
18-1200
12-3000
24-2000
12-2000
12-1000
18-2000
24-2500
1.5 cfs
4.0 cfs
2.5 cfs
3.0 cfs
2.0 cfs
5.0 cfs
1.0 cfs
I
II
III
IV
V
VI
VII
VIII
IX
X
XI
XII
XIII
XIV
XV
XVI
XVII
XIX

Solution:

Pipe No.	1	2	3	4	5	6	7
Q (cfs)	20.4	20.5	10.5	4.92	-4.16	2.76	4.29
h_f (ft)	5.58	11.31	3.03	0.70	26.58	0.23	18.77

Pipe No.	8	9	10	11	12	13	14
Q (cfs)	4.69	4.83	14.18	2.76	-4.35	-6.35	1.76
h_f (ft)	11.19	17.81	12.15	3.96	19.28	2.50	3.28

Pipe No.	15	16	17	18	19	20	21
Q (cfs)	-1.47	-6.11	-3.02	1.60	0.40	-1.06	2.56
h_f (ft)	1.74	1.06	0.22	4.11	0.10	0.93	0.43

Pipe No.	22	23	24	25	26	27	28
Q (cfs)	4.56	-4.31	2.62	1.39	3.34	3.24	3.19
h_f (ft)	0.91	14.23	10.67	1.04	8.61	16.27	7.89

Pipe No.	29	30	31	32	33	34	35
Q (cfs)	2.63	7.23	-2.73	-8.32	-10.82	-12.82	3.02
h_f (ft)	5.39	65.76	0.97	8.50	1.61	5.63	7.07

Pipe No.	36	37	38	39	40	41	42
Q (cfs)	0.98	17.83	-0.73	-8.37	12.10	4.57	12.33
h_f (ft)	0.80	8.62	0.06	6.45	3.02	1.58	13.82

Pipe No.	43	44	45	46	47	48	49
Q (cfs)	12.85	8.34	30.45	27.18	57.63	20.11	-2.61
h_f (ft)	10.01	5.12	18.60	9.90	7.77	8.19	7.08

Pipe No.	50	51
Q (cfs)	-5.16	10.55
h_f (ft)	27.05	3.07

This solution took seven iterations to meet the error requirement that the sum of flow rate changes be less than 0.01 and required 11.7 seconds of execution time on a UNIVAC 1108 computer.

9. Solve the 65 pipe network below for the flow rates in all pipes, the head losses in all pipes, the elevation of the HGL at all junctions, and the pressure at all junctions. All pipes have e = 0.0102 inch. Three values of flow rate and head loss for the five pumps in the network are:

Pump No.	Point 1		Point 2		Point 3	
	Q_p	h_p	Q_p	h_p	Q_p	h_p
1	8.	200.	11.	180.	16.	80.
2	5.	180.	7.5	150.	10.	50.
3	4.	200.	6.	180.	10.	80.
4	4.	250.	6.	210.	8.	150.
5	2.	50.	4.	40.	6.	20.

The elevations of the junctions are:

Junction No.	Elev. ft	Junction No.	Elev. ft	Junction No.	Elev. ft
1	150	11	190	21	170
2	150	12	190	22	160
3	160	13	180	23	150
4	150	14	160	24	150
5	165	15	150	25	140
6	160	16	150	26	130
7	180	17	140	27	130
8	190	18	150	28	130
9	200	19	170	29	125
10	180	20	180		

Solution:

Pipe No.	1	2	3	4	5	6	7
Q (cfs)	11.61	3.18	2.18	1.50	2.22	-1.19	0.30
h_f (ft)	82.15	14.20	9.23	23.43	26.31	2.87	0.18

Pipe No.	8	9	10	11	12	13	14
Q (cfs)	-0.68	0.49	-1.43	2.63	2.06	10.00	0.93
h_f (ft)	3.60	0.73	3.42	7.68	11.10	25.67	1.89

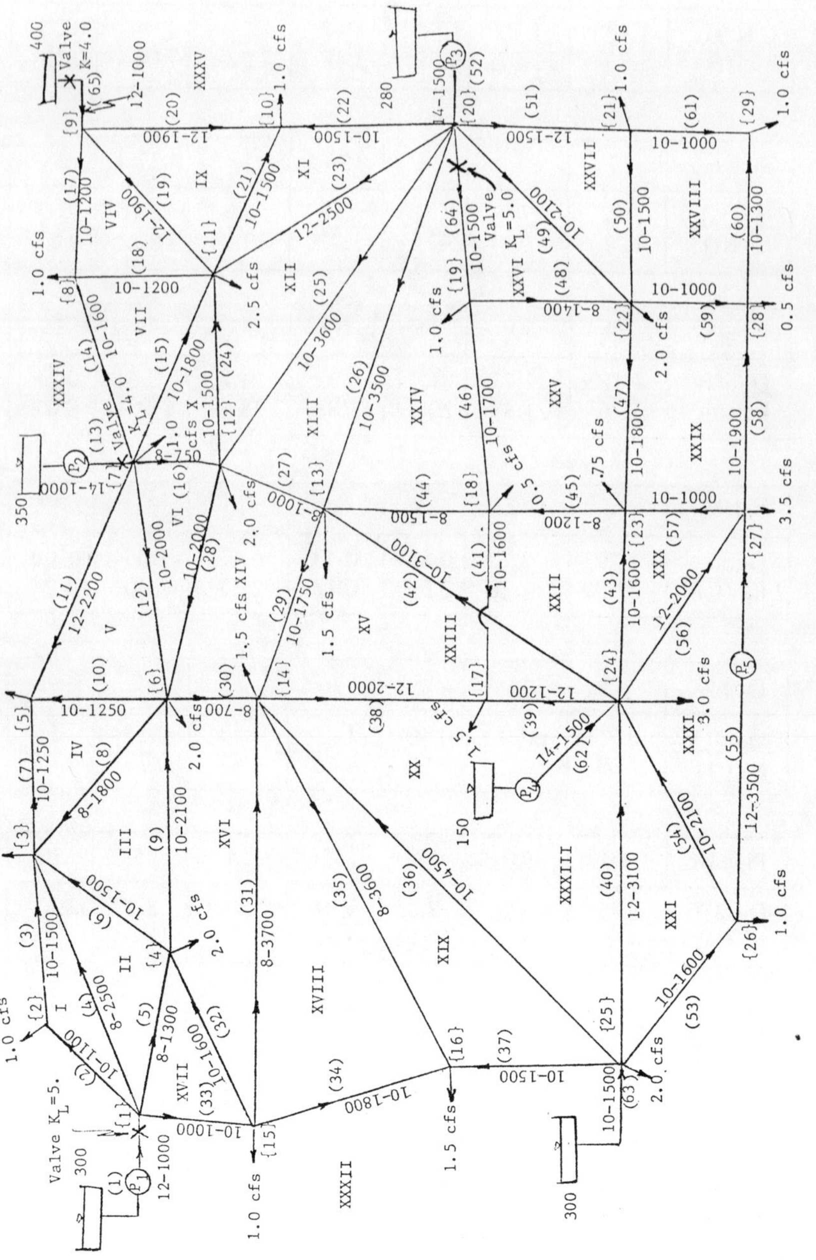

Pipe No.	15	16	17	18	19	20	21
Q (cfs) h_f (ft)	1.49 5.33	1.89 11.07	1.54 3.78	1.47 3.44	2.75 7.22	2.71 7.04	-0.26 0.18

Pipe No.	22	23	24	25	26	27	28
Q (cfs) h_f (ft)	-1.45 4.21	-1.77 4.03	-1.71 5.74	0.58 1.71	2.09 19.84	-2.10 18.13	0.07 0.02

Pipe No.	29	30	31	32	33	34	35
Q (cfs) h_f (ft)	1.18 3.26	2.73 21.37	1.14 20.22	-0.92 1.88	4.72 28.18	3.51 28.34	0.71 8.01

Pipe No.	36	37	38	39	40	41	42
Q (cfs) h_f (ft)	-1.95 22.30	-2.72 14.28	0.88 0.85	-0.43 0.13	-3.72 21.31	1.05 2.40	-1.00 4.25

Pipe No.	43	44	45	46	47	48	49
Q (cfs) h_f (ft)	-0.86 1.63	-0.50 1.72	-0.40 0.90	1.45 4.75	1.43 4.87	0.35 0.79	2.55 17.59

Pipe No.	50	51	52	53	54	55	56
Q (cfs) h_f (ft)	1.45 4.21	4.25 13.38	9.03 26.61	1.24 3.32	-3.02 24.63	3.27 18.65	-1.21 1.55

Pipe No.	57	58	59	60	61	62	63
Q (cfs) h_f (ft)	-0.22 0.08	-1.22 3.78	0.92 1.17	-0.80 1.15	1.80 4.23	6.25 12.87	-5.14 19.47

Pipe No.	64	65
Q (cfs) h_f (ft)	2.80 16.80	7.00 23.88

Junction No.	1	2	3	4	5	6
HGL (ft)	390.2	376.0	366.7	363.9	366.6	363.1
Pressure (psi)	104.1	97.9	89.6	92.7	87.3	88.0

Junction No.	7	8	9	10	11	12
HGL (ft)	374.2	372.3	376.1	369.1	368.9	363.2
Pressure (psi)	84.2	79.0	76.3	81.9	77.5	75.0

Junction No.	13	14	15	16	17	18
HGL (ft)	345.0	341.8	362.0	333.8	340.9	343.3
Pressure (psi)	71.5	78.8	91.9	79.6	87.1	83.8

Junction No.	19	20	21	22	23	24
HGL (ft)	348.1	364.9	351.5	347.3	342.4	340.8
Pressure (psi)	77.2	80.1	78.65	81.2	83.4	82.7

Junction No.	25	26	27	28	29
HGL (ft)	319.5	316.1	342.3	346.1	347.3
Pressure (psi)	77.8	80.7	92.0	93.6	96.3

Using the linear theory program described in Appendix C, the solution was obtained in six iterations with a sum of changes in flow rate equal to 0.019. The solution requires 23.3 seconds of execution time on a UNIVAC 1108.

10. Determine the flow rates and head losses of each of the 63 pipes of the network shown below. The diameter of each pipe is given in centimeters and the length of each pipe is given in meters. Thus pipe number 1 is 16 cm in diameter and has a length of 1500 m. All 63 pipes of the network have a wall roughness of 0.026 cm. The network is supplied by 2 reservoirs without pumps and 4 pumps obtaining their water from the reservoirs as shown in the sketch. A booster pump exists in pipe 43 also. The water surface elevations are as given on the sketch in meters and the pump characteristic curves are defined by the data for three points as given below.

Pump No.	Point 1		Point 2		Point 3	
	Q m^3/s	h_p(m)	Q m^3/s	h_p(m)	Q m^3/s	h_p(m)
1	0.03	84.	0.045	70.	0.06	52.
2	0.06	45.	0.105	39.	0.15	31.
3	0.06	45.	0.105	39.	0.15	31.
4	0.06	45.	0.105	39.	0.15	31.
5	0.03	9.0	0.045	7.5	0.06	5.5

The elevation of the junctions are as given below. What is the elevation of the HGL at each junction, the head in meters, and the pressure?

Junction No.	1	2	3	4	5	6	7	8	9	10
Elev. (m)	300	250	255	260	280	200	340	270	330	360

Junction No.	11	12	13	14	15	16	17	18	19	20
Elev. (m)	260	420	270	240	260	260	260	280	270	300

Junction No.	21	22	23	24	25	26	27	28	29	30
Elev. (m)	250	260	200	160	160	70	70	40	20	260

Junction No.	31	32	33
Elev. (m)	200	-50	-60

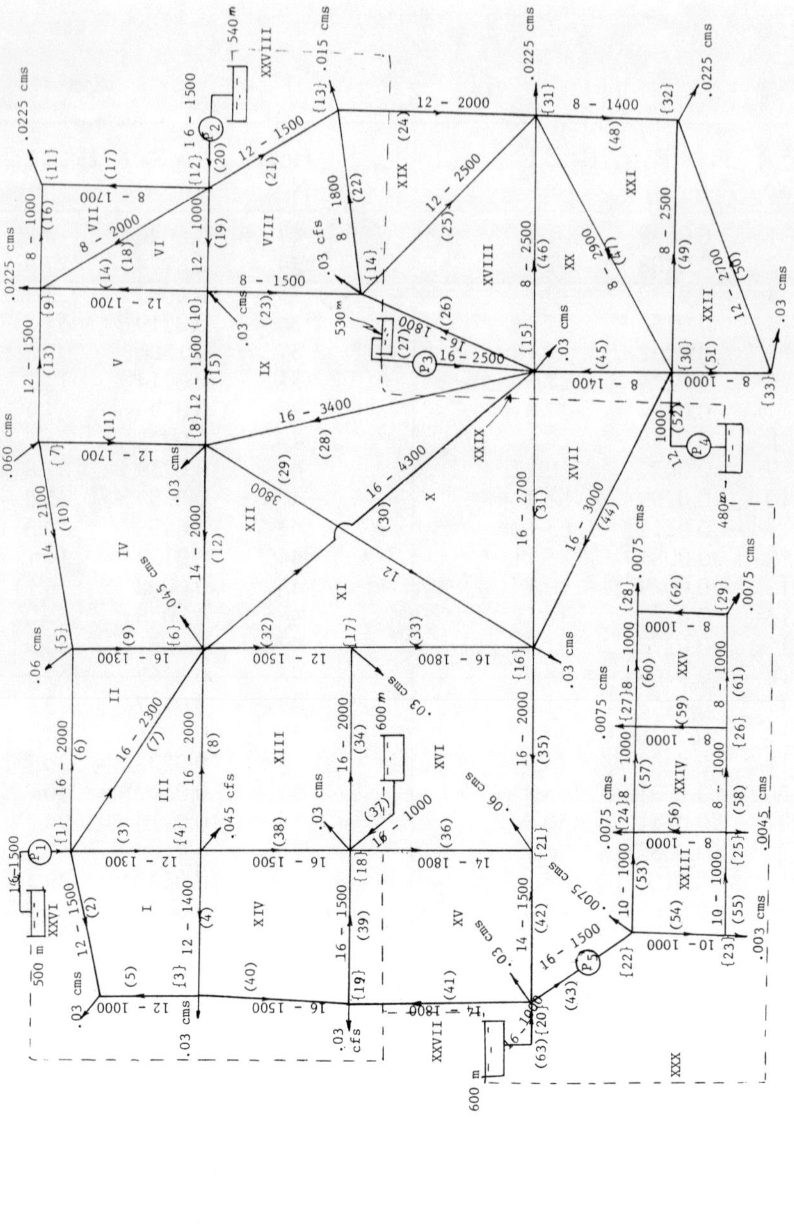

Answer

Pipe No.	Flow Rate (m^3/s)	Head Loss (m)	Pipe No.	Flow Rate (m^3/s)	Head Loss (m)
1	0.0869	201.1	33	0.0046	0.80
2	0.0208	53.6	34	0.0274	27.6
3	0.0132	19.1	35	0.0094	3.5
4	0.0153	27.4	36	0.0219	31.8
5	0.0092	7.16	37	0.1306	297.8
6	0.0274	27.5	38	-0.0214	12.7
7	0.0254	27.3	39	-0.0299	24.4
8	0.0148	8.3	40	-0.0238	15.7
9	-0.0027	0.22	41	0.0239	37.8
10	0.0299	68.4	42	0.0287	45.2
11	0.0212	63.1	43	0.0450	54.6
12	0.0081	5.11	44	0.0139	11.1
13	0.0089	10.1	45	-0.0002	0.11
14	0.0164	38.1	46	0.0056	55.0
15	0.0273	91.0	47	0.0051	54.9
16	0.0107	78.9	48	0.0167	267.5
17	0.0118	165.5	49	0.0137	322.4
18	0.0078	86.5	50	-0.0078	14.3
19	0.0243	48.4	51	0.0222	336.7
20	0.0791	166.9	52	0.0547	240.7
21	0.0351	150.0	53	0.0210	94.2
22	-0.0035	16.3	54	0.0165	58.1
23	0.0106	117.7	55	0.0135	39.1
24	0.0166	45.8	56	-0.0020	2.97
25	0.0119	29.7	57	0.0116	93.0
26	0.0277	25.3	58	0.0109	82.9
27	0.0805	287.6	59	0.0031	7.07
28	-0.0043	1.37	60	0.0072	36.2
29	0.0061	12.52	61	0.0078	43.2
30	-0.0065	3.74	62	0.0003	0.12
31	0.0148	11.16	63	0.1276	284.5
32	0.0071	6.62			

Junction	HGL (m)	Head (m)	Pressure (N/cm^2)
1	308.6	8.6	8.43
2	255.0	5.0	4.92
3	262.2	7.2	7.03
4	289.6	29.6	29.0
5	281.1	1.12	1.09
6	281.3	1.34	1.31
7	349.5	9.5	9.31
8	286.4	16.4	16.1
9	339.4	3.4	9.19
10	377.5	17.5	17.1
11	260.4	0.45	0.44
12	425.9	5.93	5.81
13	275.9	5.88	5.76
14	259.6	15.6	15.3
15	284.9	24.9	24.4
16	273.8	13.8	13.5
17	274.6	14.6	14.3
18	302.2	22.2	21.7
19	277.7	7.7	7.56
20	315.5	15.5	15.2
21	270.3	20.3	19.9
22	268.4	8.4	8.25
23	210.4	10.4	10.2
24	174.2	14.2	14.0
25	171.3	11.3	11.0
26	88.3	18.3	17.9
27	81.3	11.3	11.0
28	45.1	5.1	5.0
29	45.2	25.2	24.7
30	284.8	24.8	24.4
31	230.1	30.1	29.5
32	-37.5	12.5	12.2
33	-51.8	8.2	8.0

11. Solve the 8-pipe network shown in the accompanying sketch which is supplied by two reservoirs with water surfaces at 180 and 200 ft, respectively. The PRV is located 450 ft downstream from the upstream junction of pipe 5 and has a pressure setting which maintains the elevation of the HGL at 50 ft downstream from the valve.

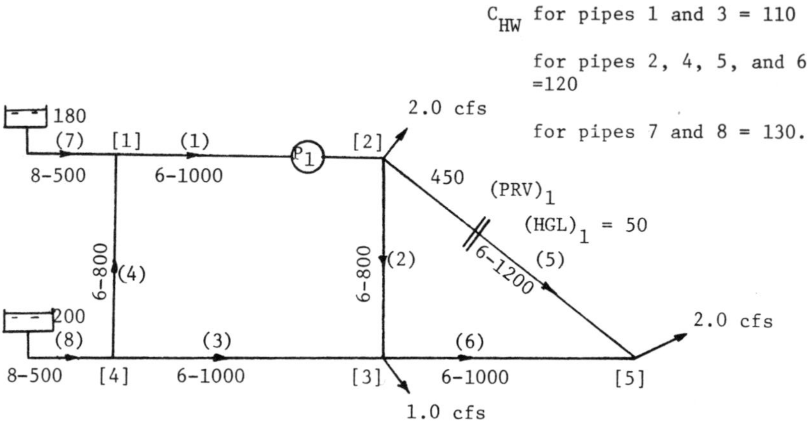

Pump Characteristics

$Q_{P_1} = 1.0$ cfs, $h_{P_1} = 40.'$, $Q_{P_2} = 1.5$ cfs, $h_{P_2} = 35.'$, $Q_{P_3} = 2.0$ cfs, $h_{P_3} = 26.'$

Solution:

Pipe No.	1	2	3	4
Flow rate (cfs) Head loss (ft)	2.56 130.61	-0.32 1.89	2.44 119.67	0.73 8.75

Pipe No.	5	6	7	8
Flow rate (cfs) Head loss (ft)	0.88 18.42	1.12 24.12	1.83 6.33	3.17 17.88

HGL upstream from PRV = 53.80 ft
HGL downstream from PRV = 50.00 ft

12. Obtain the flow rates in each pipe, and the pressure at each junction of the network in the accompanying sketch. Water is flowing (in SI units the kinematic viscosity $\nu = 0.00000113$ N²/s = 0.0133 Stokes, and the specific weight $\gamma = 9800$ N/m³).

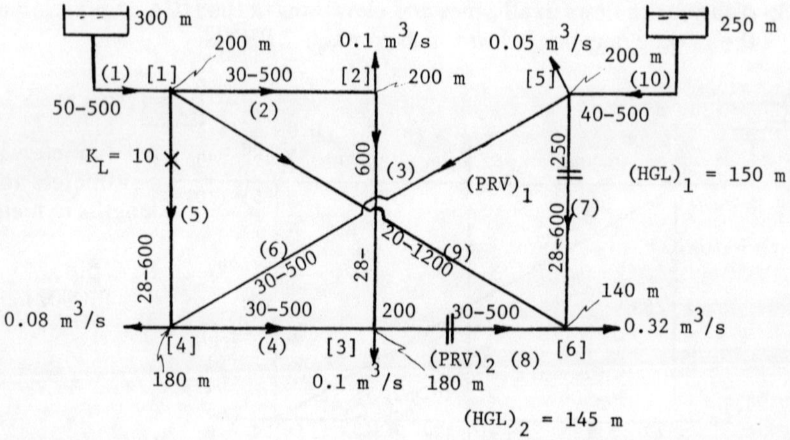

Solution:

Pipe No.	1	2	3	4	5
Flow rate m^3/s Head loss (m)	0.6402 6.00	0.2857 16.21	0.1857 12.32	-0.0857 1.79	0.2344 30.32

Pipe No.	6	7	8	9	10
Flow rate m^3/s Head loss (m)	-0.2402 11.77	0.0 0.0	0.0 0.0	0.12 56.47	-0.1902 1.91

Junction No.	1	2	3	4	5	6
Elev. of HGL Pressure (kPa)	294.0	277.79	265.48	263.68	251.91	237.54

PRV No.	1	2
Upstream, HGL (m) Downstream, HGL (m)	251.91 237.54	265.48 237.54

13. Analyze the flows in all pipes and elevations of the HGL at junction of the 10 pipe network below which contains 2 PRV's.

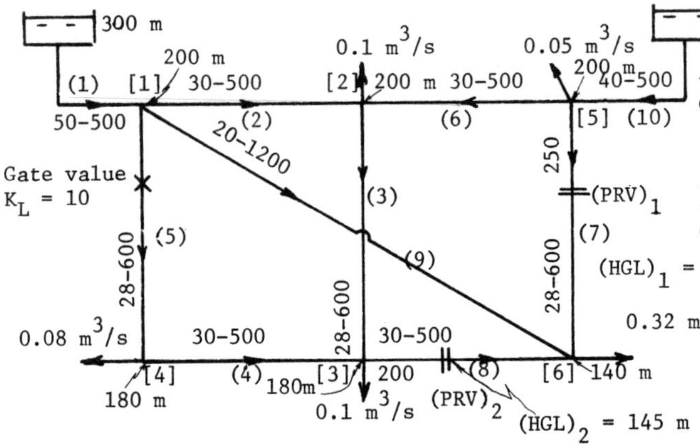

All diameters in centimeters and lengths in meters

PVC pipe
e = 0.00021 cm

Solution:

Pipe No.	1	2	3	4	5
Flow rate (m^3/s) Head loss (m)	0.7611 8.25	0.3640 25.32	-0.0169 0.16	0.1169 3.15	0.1969 22.01

Pipe No.	6	7	8	9	10
Flow rate (m^3/s) Head loss (m)	-0.2809 15.70	0.1197 5.52	0.0 0.0	0.2003 144.96	-0.1111 0.72

Junction No.	1	2	3	4	5	6
Elev. of HGL (m) Pressure (kPa)	291.745 899.1	266.422 650.9	266.583 848.5	269.737 879.4	250.718 497.0	146.781 66.5

Elevations upstream and downstream from PRV's:

PRV No.	1	2
Upstream HGL (m) Downstream HGL (m)	248.419 150.000	266.583 146.781

Note that PRV No. 2 has shut-off, whereas PRV No. 1 is permitting flow through pipe No. 7.

14. Analyze Problem No. 13 with the demand at junction 6 where pipes 7, 8, and 9 join reduced from 0.32 m^3/s to 0.12 m^3/s.

Solution:

Pipe No.	1	2	3	4	5
Flow rate (m^2/s)	0.6788	0.3626	-0.0162	0.1162	0.1962
Head loss (m)	6.68	25.15	0.15	3.12	21.87

Pipe No.	6	7	8	9	10
Flow rate (m^2/s)	-0.2788	0.0	0.0	0.1200	-0.2288
Head loss (m)	15.49	0.0	0.0	56.47	2.68

Both PRV's have shut-off with the elevation of the HGL at junction 6 equal to 236.85 m.

15. What would be the effect if lines 7 and 8 were removed from the network of Problem 13 and the demand at junction 6 increased to 0.24 m^3/s.

Solution:

The elevation of the HGL at junction 6 would drop to 88.55 m or 51.45 m below the 140 m elevation of the junction, which is impossible. Therefore, 0.24 m^3/s could not be supplied.

Chapter VI
Newton-Raphson Method

Introduction

In Chapter II the Newton-Raphson method (Eq. 2-17), one of the most widely used method for solving implicit or nonlinear equations, was described. Most books dealing with numerical methods or numerical analysis provide additional treatment of the Newton-Raphson method. It is widely used because it converges rapidly to the soltuion. In review of the discussion of the Newton-Raphson method in Chapter II, a solution to the equation $F(x) = 0$ is obtained by the iterative formula $x^{(m+1)} = x^{(m)} - F(x^{(m)})/F'(x^{(m)})$. Mathematically the convergence of the Newton-Raphson method can be examined by using Taylor's formula to evaluate $F(x) = 0$ from the function at some iterative value $x^{(m)}$; or

$$0 = F(x) = F(x^{(m)}) + (x - x^{(m)}) F'(x^{(m)}) + (x - x^{(m)})^2 F''(\xi)/2$$

in which $\xi^{(m)}$ lies between $x^{(m)}$ and x. Solving for x gives

$$x = x^{(m)} - \frac{F(x^{(m)})}{F'(x^{(m)})} - (x - x^{(m)})^2 \frac{F''(\xi)}{2F'(x^{(m)})}$$

or

$$x = x^{(m+1)} - (x - x^{(m)})^2 \frac{F''(\xi)}{2F'(x^{(m)})}$$

Thus the error of the $(m+1)^{th}$ iterate is proportional to the square of the error in the mth iterate. Convergence of this type is called quadratic convergence and in simple terms it means that each subsequent error reduction is proportional to the square of the previous error. Thus if the initial guess is 20 percent (i.e. 0.2) in error, successive iterations will produce errors of 4 percent, 1.6 percent, 0.026 percent, etc.

The Newton-Raphson method may be used to solve any of the three sets of equations describing flow in pipe networks which are discussed in Chapter IV, i.e. the equations considering (1) the flow rate in each pipe unknown, (2) the head at each junction unknown, or (3) the corrective flow rate around each loop unknown. The Newton-Raphson method requires an initial guess to the solution. Since the other two systems of equations are fewer in number than the Q-equations, the Newton-Raphson method is probably the best method to use for larger systems of equations. It requires less computer storage not only because of the fact that few simultaneous equations are included in the H- or ΔQ-equations, but also because it requires less storage for a given number of equations.

Before describing how the Newton-Raphson method can be used to solve either of the latter two systems of equations, it is necessary to extend this method from a single equation to a system of simultaneous equations. Notationally this extension is very simple. The iterative Newton-Raphson formula for a system of equations is,

$$\vec{x}^{\,(m+1)} = \vec{x}^{\,(m)} - D^{-1}\vec{F}(x^m) \quad \ldots\ldots\ldots\ldots (6\text{-}1)$$

The unknown vectors $\vec{x}$ and $\vec{F}$ replace the single varible x and function F and the inverse of the Jacobian, D^{-1}, replaces 1/dF/dx in the Newton-Raphson formula for solving a single equation. If solving the equation with the heads as the unknowns (i.e. the H-equations) the vector $\vec{x}$ becomes the vector $\vec{H}$ and if solving the equations containing the corrective loop flow rates (i.e. the ΔQ-equations) $\vec{x}$ becomes $\vec{\Delta Q}$. The individual elements for $\vec{H}$ and $\vec{\Delta Q}$ are

$$\vec{H} = \begin{vmatrix} H_1 \\ H_2 \\ \cdot \\ \cdot \\ \cdot \\ H \end{vmatrix} \begin{matrix} \text{with the known} \\ \text{H omitted from} \\ \text{the vector} \\ \\ \\ \\ \end{matrix} \quad \text{or} \quad \vec{\Delta Q} = \begin{vmatrix} \Delta Q_1 \\ \Delta Q_2 \\ \cdot \\ \cdot \\ \cdot \\ \Delta Q_L \end{vmatrix}$$

The Jacobian matrix D consists of derivative elements, individual rows of which are derivatives of that particular functional equation with respect to the variables making up the column headings. For the head equation the Jacobian is,

$$D = \begin{bmatrix} \dfrac{\partial F_1}{\partial H_1} & \dfrac{\partial F_1}{\partial H_2} & \cdots & \dfrac{\partial F_1}{\partial H_J} \\ \dfrac{\partial F_2}{\partial H_1} & \dfrac{\partial F_2}{\partial H_2} & \cdots & \dfrac{\partial F_2}{\partial H_J} \\ \vdots & & & \\ \dfrac{\partial F_J}{\partial H_1} & \dfrac{\partial F_J}{\partial H_2} & \cdots & \dfrac{\partial F_J}{\partial H_J} \end{bmatrix}$$

in which the row and column corresponding to the known head are omitted.

The last term $D^{-1}F$ in Eq. 6-1 contains the inverse of D, since division by a matrix is undefined. However in application of the Newton-Raphson method the inverse is never obtained and premultiplied by $\vec{F}$ as Eq. 6-1 implies. Rather the solution vector $\vec{z}$ of the linear system $D\vec{z} = \vec{F}$ is subtracted from the previous iterative vector of unknowns. Selecting the H-equations in the following notation, the Newton-Raphson iterative formula in practice becomes

$$\vec{H}^{(m+1)} = \vec{H}^{(m)} - \vec{z}^{(m)} \quad \ldots\ldots\ldots\ldots\ldots\ldots (6\text{-}2)$$

The equivalence of Eqs. 6-2 and 6-1 is evident since $\vec{z} = D^{-1}\vec{F}$. Since fewer computations are needed to solve the linear system $D\vec{z} = \vec{F}$ than to find the inverse D^{-1} obviously Eq. 6-2 is the form of the Newton-Raphson method used in practice. The Newton-Raphson method, therefore, obtains the solution to a system of nonlinear equations by iteratively solving a system of linear equations. In this sense it is similar to the linear theory method and can call on the same algorithm for solving a linear system of equations as does the linear theory method. It turns out, however, that the Jacobian is a symmetric matrix, and consequently an algorithm for solving a linear system of equations with a symmetric matrix might preferably be used for greater computational efficiency. The Newton-Raphson method does require a reasonably accurate initialization or it may not converge.

Head-equation

The Newton-Raphson method will be illustrated in detail by using it to solve the H-equations for the simple one loop network shown below. To

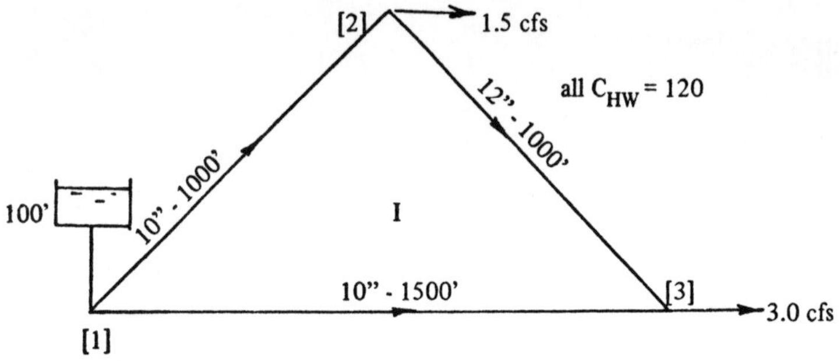

simplify the problem the Hazen-Williams equation will be used so that K and n in the exponential formula are constant. The values of K for the three pipes are: $K_{12} = 1.622$, $K_{23} = 0.667$, $K_{13} = 2.432$. The heat at junction 1 is known and equal to 100 ft. The heads H_2 and H_3 at junctions 2 and 3 are unknown and to be determined. To determine these two unknowns the H-equations will be written at junctions 2 and 3 (see Eq. 4-17 for the nature of these equations), giving

$$F_2 = -\left(\frac{H_1 - H_2}{K_{12}}\right)^{1/n_{12}} + \left(\frac{H_2 - H_3}{K_{23}}\right)^{1/n_{23}} + 1.5 = 0$$

$$F_3 = -\left(\frac{H_2 - H_3}{K_{23}}\right)^{1/n_{23}} - \left(\frac{H_1 - H_3}{K_{13}}\right)^{1/n_{13}} + 3.0 = 0$$

The equation at junction 1 is not written since $H_1 = 100$ ft is known, but might have been used in place of one of the above equations. Upon substituting known values for H_1 and the K's and n's those equations become:

$$F_2 = -\left(\frac{100 - H_2}{1.622}\right)^{0.54} + \left(\frac{H_2 - H_3}{0.667}\right)^{0.54} + 1.5 = 0$$

$$F_3 = -\left(\frac{H_2 - H_3}{0.667}\right)^{0.54} - \left(\frac{100 - H_3}{2.432}\right)^{0.54} + 3.0 = 0$$

$$\text{The Jacobian } D = \begin{vmatrix} \frac{\partial F_2}{\partial H_2} & \frac{\partial F_2}{\partial H_3} \\ \frac{\partial F_3}{\partial H_2} & \frac{\partial F_3}{\partial H_3} \end{vmatrix}$$

has the following elements

$$\frac{\partial F_2}{\partial H_2} = \frac{0.54}{K_{12}}\left(\frac{H_1 - H_2}{K_{12}}\right)^{\frac{1}{n_{12}}-1} + \frac{0.54}{K_{23}}\left(\frac{H_2 - H_3}{K_{23}}\right)^{\frac{1}{n_{23}}-1}$$

$$= 0.333\left(\frac{100 - H_2}{1.622}\right)^{-0.46} + 0.809\left(\frac{H_2 - H_3}{0.667}\right)^{-0.46}$$

$$\frac{\partial F_2}{\partial H_3} = -\frac{0.54}{K_{23}}\left(\frac{H_2 - H_3}{K_{23}}\right)^{\frac{1}{n_{23}}-1} = -0.809\left(\frac{H_2 - H_3}{0.667}\right)^{-0.46}$$

$$\frac{\partial F_3}{\partial H_2} = -\frac{0.54}{K_{23}}\left(\frac{H_2 - H_3}{K_{23}}\right)^{\frac{1}{n_{23}}-1} = -0.809\left(\frac{H_2 - H_3}{0.667}\right)^{-0.46}$$

$$\frac{\partial F_3}{\partial H_3} = \frac{0.54}{K_{23}}\left(\frac{H_2 - H_3}{K_{23}}\right)^{\frac{1}{n_{23}}-1} + \frac{0.54}{K_{13}}\left(\frac{H_1 - H_3}{K_{13}}\right)^{\frac{1}{n_{13}}-1}$$

$$= 0.809\left(\frac{H_2 - H_3}{0.667}\right)^{-0.46} + 0.222\left(\frac{100 - H_3}{2.432}\right)^{-0.46}$$

If the initialization $\vec{H} = \begin{vmatrix} H_2 \\ H_3 \end{vmatrix} = \begin{vmatrix} 95 \\ 85 \end{vmatrix}$ is used by the Newton-Raphson equation, Eq. 6-2, the solution to

$$\begin{vmatrix} 0.432 & -0.233 \\ -0.233 & 0.329 \end{vmatrix} \begin{vmatrix} z_2 \\ z_3 \end{vmatrix} = \begin{vmatrix} 3.98 \\ -3.98 \end{vmatrix}$$

is $z_2 = 4.34$ and $z_3 = -9.04$. When these are substracted from the initial guesses $H_2 = 90.66$, $H_3 = 94.04$. After completing six iterations the solution is: $H_2 = 91.45$ ft and $H_3 = 90.84$ ft. Using these heads the flow rates are computed as: $Q_{12} = 2.454$ cfs, $Q_{23} = 0.954$ cfs, and $Q_{13} = 2.046$ cfs.

A simple version of a FORTRAN computer program for solving the H-equations by the Newton-Raphson method is listed below.

```
      INTEGER N1(50),N2(50),NN(45),JB(45,7),JC(45)
      REAL H(45),D(50),Q(50),CHW(50),QJ(45),K(45),F(45,46), V(2)
C     NP-NO. PIPES, NJ-NO. OF JUNCTIONS, KNOWN-
C     JUNCTION NO. OF KNOWN HEAD, MAX-MAX. NO. OF
C     JUNCTIONS ALLOWED, ERR-ERROR PARAMETER
   98 READ(5,100,END=99) NP,NJ,KNOWN,MAX,ERR
  100 FORMAT(4I5,4F10.5)
      DO 2 I=1,NP
C     N1(I) JUNCTION NO. FROM WHICH FLOW IN PIPE COMES
C     N2(I) JUNCTION NO. TO WHICH FLOW IN PIPE GOES
C     D(I)-DIAMETER OF PIPE IN INCHES
C     CHW(I)-HAZEN-WILLIAMS COEFFICIENT FOR PIPE
C     L-LENGTH OF PIPE IN FEET
      READ(5,101) N1(I),N2(I),D(I),CHW(I),L(I)
      D(I)=D(I)/12.
    2 K(I)=4.727328*L(I)/(CHW(I)**1.85185185*D(I)**4.87037)
  101 FORMAT(2I5,5F10.5)
      NJM=NJ-1
      DO 4 J=1,NJ
C     I-JUNCTION NO.
```

```
C  QJ(I)–EXTERNAL FLOW AT JUNCTION, MINUS IF OUT
C  FROM NETWORK
C  H(I)–ESTIMATE OF HEAD AT JUNCTION USED TO
C  INITIALIZE N-R SOLUTION
     4 READ(5,102) I,QJ(I),H(I)
   102 FORMAT(I5,5F10.5)
       DO 5 J=1,NJ
       NNP=0
       DO 6 I=1,NP
       IF(N1(I) .NE. J) GO TO 7
       NNP=NNP+1
       JB(J,NNP)=I
       GO TO 6
     7 IF(N2(I) .NE. J) GO TO 6
       NNP=NNP+1
       JB(J,NNP)=-I
     6 CONTINUE
     5 NN(J)=NNP
       NCT=0
    20 SUM=0.
       JE=0
       DO 10 J=1,NJ
       IF(J .EQ. KNOWN) GO TO 10
       JE=JE+1
       JJE=J-JE
       DO 15 JJ=1,NJ
    15 F(JE,JJ)=0.
       NNP=NN(J)
       DO 11 KK=1,NNP
       II=JB(J,KK)
       I=ABS(II)
       I1=N1(I)
       I2=N2(I)
       ARG=(H(I1)-H(I2))/K(I)
       FAC=II/I
       FAC5=.54*FAC
       ARGE=ARG**.54
    13 F(JE,NJ)=F(JE,NJ)+ARGE*FAC
       IF(I1 .EQ. KNOWN) GO TO 14
       IF(I1 .GT. KNOWN) I1=I1-1
       F(JE,I1)=F(JE,I1)+FAC5*ARGE/(K(I)*ARG)
    14 IF(I2 .EQ. KNOWN) GO TO 11
       IF(I2 .GT. KNOWN) I2=I2-1
       F(JE,I2)=F(JE,I2)-FAC5*ARGE/(K(I)*ARG)
    11 CONTINUE
       F(JE,NJ)=F(JE,NJ)-QJ(J)
    10 CONTINUE
       V(1)=4.
       CALL GJR(F,46,45,N,JM,NJ,$97,JC,V)
       JE=0
       DO 24 J=1,NJ
       IF(J .EQ. KNOWN) GO TO 24
       JE=JE+1
       DIF=F(JE,NJ)
       SUM=SUM+ABS(DIF)
       H(J)=H(J)-DIF
```

```
   24 CONTINUE
      DO 25 I=1,NP
      I1=N1(I)
      I2=N2(I)
      IF(H(I1) .GT. H(I2)) GO TO 25
      WRITE(6,225) I,I1,I2
  225 FORMAT(' FLOW HAS REVERSED IN PIPE',3I5)
      N1(I)=I2
      N2(I)=I1
      II=I1
   28 NNP=NN(II)
      DO 26 KK=1,NNP
      IF(IABS(JB(II,KK)) .NE. I) GO TO 26
      JB(II,KK)=-JB(II,KK)
      GO TO 27
   26 CONTINUE
   27 IF(II .EQ. I2) GO TO 25
      II=I2
      GO TO 28
   25 CONTINUE
      NCT=NCT+1
      WRITE(6,108) NCT,SUM
  108 FORMAT(' NCT=',I5,' SUM=',E12.5)
      IF(NCT .LT. MAX .AND. SUM .GT. ERR) GO TO 20
      WRITE(6,103)(H(J),J=1,NJ)
  103 FORMAT(' HEADS AT JUNCTIONS',/,(1H ,13F10.3))
      WRITE(6,104)
  104 FORMAT(' FROM      TO    DIAMETER LENGTH
     $CHW    FLOWRATE HEAD LOSS   HEADS AT
     $JUNCTIONS')
      DO 17 I=1,NP
      I1=N1(I)
      I2=N2(I)
      DH=H(I1)-H(I2)
      Q(I)=(DH/K(I))**.54
   19 WRITE(6,105) I1,I2,D(I),L(I),CHW(I),Q(I),DH,H(I1),H(I2)
  105 FORMAT(2I5,2F10.1,F10.0,4F10.3)
   17 CONTINUE
      GO TO 98
   97 WRITE(6,306) JC(1),V
  306 FORMAT(' OVERFLOW OCCURRED-CHECK SPEC.
     $FOR REDUNDANT EQ. RESULTING IN SINGULAR
     $MATRIX' ,I5,2F10.2)
   99 STOP
      END
```

Example Problems Based on the H-Equations

1. Solve the network below for the heads at each junction. From these computed heads compute the flow rates in each pipe. The head at junction 3 is to be maintained at 100 ft.

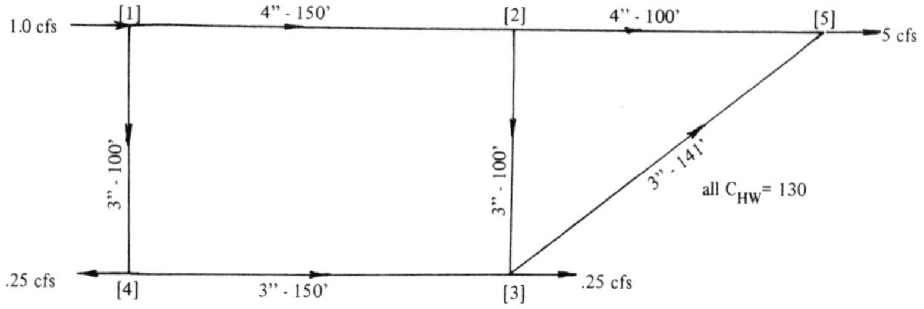

Solution:

The system of equations for this network is:

$$F_1 = \left(\frac{H_1 - H_2}{18.19}\right)^{0.54} + \left(\frac{H_1 - H_4}{49.22}\right)^{0.54} - 1.0 = 0$$

$$F_2 = -\left(\frac{H_1 - H_2}{18.19}\right)^{0.54} + \left(\frac{H_2 - H_5}{49.22}\right)^{0.54} + \left(\frac{H_2 - H_3}{12.12}\right)^{0.54} = 0$$

$$F_4 = -\left(\frac{H_1 - H_4}{49.22}\right)^{0.54} + \left(\frac{H_4 - 100}{73.833}\right) + 0.25 = 0$$

$$F_5 = -\left(\frac{H_2 - H_5}{12.12}\right)^{0.54} - \left(\frac{100 - H_5}{69.40}\right)^{0.54} + 0.5 = 0$$

If as initialization, $H_1 = 115'$, $H_2 = 103'$, $H_4 = 103'$, and $H_5 = 99'$, then

$$D\vec{z} = \vec{F} \text{ becomes } \begin{bmatrix} 0.057 & -0.036 & -0.021 & 0.0 \\ -0.036 & -0.150 & 0.0 & -0.074 \\ -0.021 & 0.0 & 0.53 & 0.0 \\ 0.0 & -0.074 & 0.0 & -0.129 \end{bmatrix} \begin{bmatrix} z_1 \\ z_2 \\ z_4 \\ z_5 \end{bmatrix} = \begin{bmatrix} 0.266 \\ -0.029 \\ -0.039 \\ -0.151 \end{bmatrix}$$

A solution of which gives: $z_1 = 5.78$, $z_2 = 0.86$, $z_4 = 1.55$, and $z_5 = -0.67$ producing $H_1 = 109.22$, $H_2 = 102.14$, $H_4 = 101.45$, $H_5 = 99.67$. After two additional iterations: $H_1 = 109.74'$, $H_2 = 102.19'$, $H_3 = 100'$, $H_4 = 101.63'$, $H_5 = 99.58'$, and $Q_{12} = 0.622$ cfs, $Q_{14} = 0.378$ cfs, $Q_{43} = 0.128$ cfs, $Q_{25} = 0.436$ cfs, $Q_{23} = 0.186$ cfs, $Q_{35} = 0.064$ cfs.

2. Obtain the elevation of the HGL at each junction and the flow in each pipe in the network below

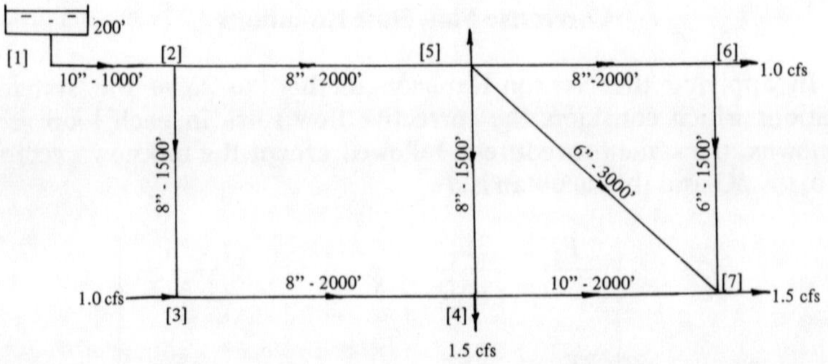

Solution:

The system of equations is:

$$F_2 = -\left(\frac{200 - H_2}{1.622}\right)^{0.54} + \left(\frac{H_2 - H_5}{9.615}\right)^{0.54} + \left(\frac{H_2 - H_3}{7.211}\right)^{0.54} = 0$$

$$F_3 = -\left(\frac{H_2 - H_3}{7.211}\right)^{0.54} + \left(\frac{H_3 - H_4}{9.615}\right)^{0.54} - 1.0 = 0$$

$$F_4 = -\left(\frac{H_3 - H_4}{9.615}\right)^{0.54} - \left(\frac{H_5 - H_4}{7.211}\right)^{0.54} + \left(\frac{H_4 - H_7}{3.243}\right)^{0.54} + 0.5 = 0$$

$$F_5 = -\left(\frac{H_2 - H_5}{9.615}\right)^{0.54} + \left(\frac{H_5 - H_6}{9.615}\right)^{0.54} + \left(\frac{H_5 - H_4}{7.211}\right)^{0.54} + \left(\frac{H_5 - H_7}{58.55}\right)^{0.54} + 0.5 = 0$$

$$F_6 = -\left(\frac{H_5 - H_6}{9.615}\right)^{0.54} + \left(\frac{H_6 - H_7}{29.27}\right)^{0.54} + 1.0 = 0$$

$$F_7 = -\left(\frac{H_6 - H_7}{3.243}\right)^{0.54} - \left(\frac{H_5 - H_7}{58.55}\right)^{0.54} - \left(\frac{H_4 - H_7}{3.243}\right)^{0.54} + 1.5 = 0$$

Solving these equations by the Newton-Raphson method gives: H_2 = 183.5', H_3 = 174.1', H_4 = 134.2', H_5 = 137.0', H_6 = 128.9', H_7 = 129.2', and the flows are: Q_{12} = 3.5, Q_{25} = 2.344, Q_{23} = 1.156, Q_{34} = 2.156, Q_{54} = 0.599, Q_{57} = 0.335, Q_{56} = 0.910, Q_{47} = 1.255, Q_{76} = 0.090 all in cfs.

Corrective Flow Rate Equations

In applying the Newton-Raphson method to solve the system of equations which considers the corrective flow rates in each loop as the unknowns, the same procedure is followed except the unknown vector in Eq. 6-1 is ΔQ and the Jacobian is,

$$D = \begin{vmatrix} \frac{\partial F_1}{\partial \Delta Q_1} & \frac{\partial F_1}{\partial \Delta Q_2} & \cdots & \frac{\partial F_L}{\partial \Delta Q_L} \\ \frac{\partial F_2}{\partial \Delta Q_1} & \frac{\partial F_2}{\partial \Delta Q_2} & \cdots & \frac{\partial F_L}{\partial \Delta Q_L} \\ \vdots & \vdots & & \vdots \\ \frac{\partial F_L}{\partial \Delta Q_1} & \frac{\partial F_L}{\partial \Delta Q_2} & \cdots & \frac{\partial F_L}{\partial \Delta Q_L} \end{vmatrix}$$

With $\vec{z}$ defined as the solution to $D^{(m)}\vec{z}^{(m)} = \vec{F}^{(m)}$ as previous where F now becomes the equations evaluated from the mth iterative values of $\Delta Q^{(m)}$, the Newton-Raphson method becomes

$$\Delta Q^{(m+1)} = \Delta Q^{(m)} - \vec{z} \quad \ldots\ldots\ldots\ldots\ldots\ldots (6\text{-}3)$$

The Newton-Raphson method will be illustrated in solving the ΔQ-equations by giving the details involved in solving the two loop network shown below. The table to the right of the sketch contains values of a possible initial flow rate in each pipe of the network which satisfy the junction continuity equations and which will be used to define the corrective loop flow rate equations. To simplify the illustration the Hazen-Williams formula will be used.

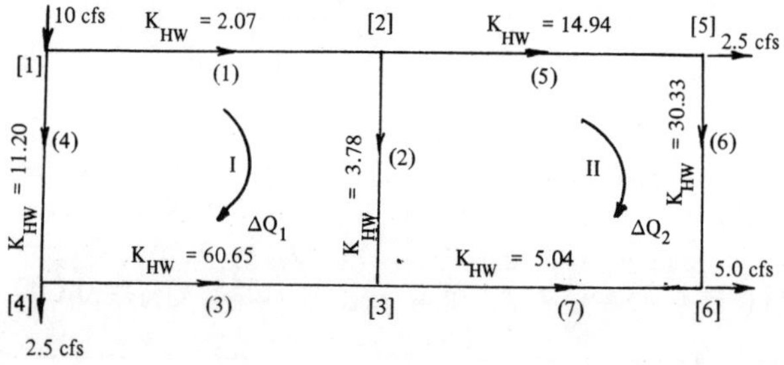

Pipe No.	Q_o(cfs)
1	5.0
2	2.0
3	2.5
4	5.0
5	3.0
6	0.5
7	4.5

Since there are two loops, there are two corrective flow rates, ΔQ_1 and ΔQ_2 which are unknown. Writing the energy equation around these two loops (with head losses in the clockwise direction as positive), gives the following two simultaneous equations to solve for these two unknowns.

$$F_1 = 2.07\,(5 + \Delta Q_1)^{1.85} + 3.78\,(2 + \Delta Q_1 - \Delta Q_2)^{1.85}$$
$$- 60.65\,(2.5 - \Delta Q_1)^{1.85} - 11.20\,(5 - \Delta Q_1)^{1.85} = 0$$

$$F_2 = 14.94\,(3 + \Delta Q_2)^{1.85} + 30.33\,(.5 + \Delta Q_2)^{1.85}$$
$$- 5.04\,(4.5 - \Delta Q_2)^{1.85} - 3.78\,(-2 - \Delta Q_2 + \Delta Q_1)^{1.85} = 0$$

The four elements of the Jacobian are:

$$\frac{\partial F_1}{\partial \Delta Q_1} = 3.83\,(5 + \Delta Q_1)^{0.85} + 6.99\,(2 + \Delta Q_1 - \Delta Q_2)^{0.85}$$
$$+ 112.20\,(2.5 + \Delta Q_1)^{0.85} + 20.72\,(5 + \Delta Q_1)^{0.85}$$

$$\frac{\partial F_1}{\partial \Delta Q_2} = -6.99\,(2 + \Delta Q_1 - \Delta Q_2)^{0.85}$$

$$\frac{\partial F_2}{\partial \Delta Q_1} = -6.99\,(2 + \Delta Q_2 - \Delta Q_1)^{0.85}$$

$$\frac{\partial F_2}{\partial \Delta Q_2} = 27.64\,(3 + \Delta Q_2)^{0.85} + 56.11\,(.5 + \Delta Q_2)^{0.85}$$
$$+ 9.32\,(4.5 - \Delta Q_2)^{0.85} + 6.99\,(2 - \Delta Q_2 + \Delta Q_1)^{0.85}$$

Note that the Jacobian is a symmetric matrix as was the Jacobian from the H-equations.

Starting the Newton iteration with $\Delta Q_1 = \Delta Q_2 = 0$, then

$$\begin{bmatrix} 353.5 & -12.6 \\ -12.6 & 147.5 \end{bmatrix} \begin{bmatrix} z_1 \\ z_2 \end{bmatrix} = \begin{bmatrix} -497. \\ 27.4 \end{bmatrix}$$

results upon evaluating the functions F_1 and F_2 and the elements in the Jacobian. Solution of this system produces $z_1 = -\Delta Q_1^{(1)} = -1.399$, $z_2 = -\Delta Q_2^{(0)} = 0.0654$. For the second Newton-Raphson iteration,

$$\begin{bmatrix} 222.5 & -20.2 \\ -20.2 & 151.1 \end{bmatrix} \begin{bmatrix} z_1 \\ z_2 \end{bmatrix} = \begin{bmatrix} 90.3 \\ -5.46 \end{bmatrix}$$

which gives $z_1 = -0.414$, $z_2 = -0.0914$, and $\Delta Q_1^{(2)} = 1.813$ and $\Delta Q_2^{(2)} = 0.026$. After two additional iterations changes in corrective flow rates are insignificant and the solution is accepted as $\Delta Q_1 = 1.866$ and $\Delta Q_2 = 0.0331$. The flow rates in each pipe can now be computed by adding these corrective flow rates to the initially assumed values which satisfy the junction continuity equations. From these flow rates, the frictional head losses can be computed. These results are:

Pipe No.	1	2	3	4	5	6	7
Q (cfs)	6.87	3.83	0.634	3.134	3.03	0.533	4.47
h_f (ft)	73.5	45.5	26.1	92.9	116.6	9.46	80.6

A listing is given below of a simplifed FORTRAN computer program for carrying out the computations described above for solving the ΔQ-equations. Below this listing is a listing of the input data required to solve example problem 1 below by this program.

FORTRAN program for solving the corrective loop flow rate equation by the Newton-Raphson method for analyzing pipe networks.

```
      REAL D(50),L(50),K(50),CHW(50),QI(50),DQ(50),DR(50,
     $51),V(2),A(3),B(3),HO(3),DELEV(3)
      INTEGER LP(42,7),NN(42),LO(3),LLP(3),LOP(50,4),
     $NLOP(50)
C     NP-NO. OF PIPES, NL-NO. OF LOOPS, MAX-MAX. NO.
C     OF ITERATIONS ALLOWED, NPUMP-NO. OF PUMPS,
C     NSL-NO. OF PSEUDO LOOPS.
   98 READ(5,100,END=99) NP,NL,MAX,NPUMP,NSL
      NLP=NL+1
  100 FORMAT(10I5)
C     II-PIPE NO., D(II)-DIAMETER OF PIPE IN INCHES, L(II)-
C     LENGTH OF PIPE IN FT, CHW(II)-HAZEN WILLIAMS COEF.,
C     QI(II)-INITIAL FLOW SATISFYING CONTINUITY EQS.
      DO 1 I=1,NP
      READ(5,101) II,D(II),L(II),CHW(II),QI(II)
      D(II)=D(II)/12.
    1 K(II)=4.77*L(II)/(D(II)**4.87*CHW(II)**1.852)
  101 FORMAT(I5,7F10.5)
      DO 2 I=1,NL
      DQ(I)=0.
C     NNP IS THE NUMBER OF PIPES AROUND THE LOOP
C     LP(I,J) ARE THE PIPE NO. AROUND THE LOOP. IF
C     COUNTERCLOCKWISE THIS NO. IS -
      READ(5,100) NNP,(LP(I,J),J=1,NNP)
    2 NN(I)=NNP
C     LLP(I)-LINE NO. CONTAINING PUMP (MINUS IF
C     COUNTERCLOCKWISE, A, B, HO-PUMP CHAR
      IF(NPUMP .EQ. 0) GO TO 30
      DO 31 I=1,NPUMP
   31 READ(5,101)  LLP(I),A(I),B(I),HO(I)
C     LO(I)-NO. OF PSEUDO LOOP, DELEV(I)-ELEV. DIFF. ON
C     RIGHT OF = IN ENERGY EQ.
      DO 32 I=1,NSL
   32 READ(5,101) LO(I),DELEV(I)
   30 DO 50 I=1,NP
      NLO=0
```

```
      DO 51 L1=1,NL
      NNP=NN(L1)
      DO 51 KK=1,NNP
      IF (IABS(LP(L1,KK)) .NE. I) GO TO 51
      NLO=NLO+1
      LOP(I,NLO)=L1*LP(L1,KK)/I
   51 CONTINUE
   50 NLOP(I)=NLO
      NCT=0
   10 SUM=0.
      DO 3 I=1,NL
      DO 12 J=1,NLP
   12 DR(I,J)=0.
      NNP=NN(I)
      DO 3 J=1,NNP
      IJ=LP(I,J)
      IIJ=IABS(IJ)
      Q=QI(IIJ)
      NLO=NLOP(IIJ)
      DO 52 KK=1,NLO
      L1=LOP(IIJ,KK)
      LL=IABS(L1)
   52 Q=Q+FLOAT(L1/LL)*DQ(LL)
      QE=ABS(Q)**.852
      FAC=IJ/IIJ
      DR(I,NLP)=DR(I,NLP)+FAC*K(IIJ)*Q*QE
      DO 53 KK=1,NLO
      L1=LOP(IIJ,KK)
      LL=IABS(L1)
   53 DR(I,LL)=DR(I,LL)+FAC*FLOAT(L1/LL)*1.852*K(IIJ)*QE
    3 CONTINUE
      IF(NSL .EQ. 0) GO TO 40
      DO 33 I=1,NSL
      II=LO(I)
      DR(II,NLP)=DR(II,NLP)-DELEV(I)
      NNP=NN(II)
      DO 33 IK=1,NPUMP
      IL=IABS(LLP(IK))
      DO 33 KK=1,NNP
      IF(IL .NE. IABS(LP(II,KK))) GO TO 33
      Q=ABS(FLOAT(LLP(IK)/IL)*QI(IL)+DQ(II)
      HP=(A(IK)*Q+B(IK))*Q+HO(IK)
      IF(LLP(IK) .LT. 0) GO TO 35
      DR(II,NLP)=DR(II,NLP)-HP
      DR(II,II )=DR(II,II )+2.*A(IK)*Q+B(IK)
      GO TO 33
   35 DR(II,NLP)=DR(II,NLP)+HP
      DR(II,II )=DR(II,II )-2.*A(IK)*Q+B(IK)
   33 CONTINUE
   40 V(1)=4.
      CALL GJR(DR,51,50,NL,NLP,$98,D,V)
      DO 7 I=1,NL
      SUM=SUM+ABS(DR(I,NLP))
    7 DQ(I)=DQ(I)-DR(I,NLP)
      NCT=NCT+1
      WRITE(6,202) NCT,SUM,(DQ(I),I=1,NL)
```

```
202 FORMAT(' NCT='I2,(12F10.3))
    IF(SUM .GT. .001 .AND. NCT .LT. MAX) GO TO 10
    IF(NCT .EQ. MAX) WRITE(6,102) NCT,SUM
102 FORMAT(' DID NOT CONVERGE --NCT=',I5,' SUM=',
   $E12.5)
    DO 8 I=1,NL
    NNP=NN(I)
    DO 8 J=1,NNP
    IJ=LP(I,J)
    IIJ=IABS(IJ)
  8 QI(IIJ)=QI(IIJ)+FLOAT(IJ/IIJ)*DQ(I)
    WRITE(6,105) (QI(I),I=1,NP)
105 FORMAT(' FLOWRATES IN PIPES',/,(1H ,13F10.3))
    DO 9 I=1,NP
  9 D(I)=K(I)*ABS(QI(I))**1.852
    WRITE(6,106) (D(I),I=1,NP)
106 FORMAT(' HEAD LOSSES IN PIPES',/,(1H ,13F10.3))
    GO TO 98
 99 STOP
    END
```

The input data cards needed in order to use the previously listed program follow:

16	7	15	1	1		
1	12.		1000.	130.	2.5	Pipe diameters, lengths, and coef. C_{HW} and initial flow rate Q_o
2	8.		1000.	130.	1.0	
3	10.		1200.	130.	.75	
4	10.		900.	130.	2.0	
5	10.		1000.	130.	.5	
6	10.		2500.	130.	1.0	
7	8.		1000.	130.	.25	
8	10.		1400.	130.	.75	
9	12.		2000.	130.	.5	
10	8.		1200.	130.	2.	
11	8.		1300.	130.	1.	
12	10.		1600.	130.	.5	
13	10.		1000.	130.	1.	
14	10.		1300.	130.	1.	
15	12.		900.	130.	2.25	
16	12.		500.	130.	2.5	
4	2	3	-4	-5		Pipe Nos. in Loops
4	-7	8	-15	-3		
4	2	-7	8	-6		
3	11	12	-9			
3	9	-10	-8			
3	10	13	-14			
4	16	7	-2	-1		
-1	-2.505		16.707	155.286		Pump characteristics
7	150.					Pseudo Loop No. & Δ Elev.

Input data needed for example problem 1 which follows:

8	3	10			
1	12.	2000.	95.	5.	
2	10.	1500.	95.	2.	
3	6.	2000.	95.	2.5	
4	8.	1500.	95.	5.	
5	8.	2000.	95.	3.	Pipe Information
6	6.	1000.	95.	.5	
7	10.	2000.	95.	4.5	
8	10.	3000.	95.	2.	
3	1	2	-8		
4	5	6	-7	-2	Pipes in Loop
3	8	-3	-4		

The input data needed for a more extensive computer program using this method is given in Appendix C. This latter computer program permits pumps, reservoirs, etc., to exist in the network does not require information about pipe numbers in each loop, generates its own initialization and for large networks is efficient in computer time and storage since it orders loops in such a manner so that all nonzero elements of the Jacobian are concentrated in a band near the diagonal elements.

Including Pressure Reducing Valves in Analyses Based on the ΔQ Equations

An important fact which allows pipe networks to be analyzed by the system of ΔQ equations is that junction continuity is satisfied regardless of what values each corrective loop flow rate ΔQ takes on provided the initializing flows, Q_{oi}, satisfy all junction continuity equations. Thus after establishing initial flows which satisfy continuity at each junction no additional equations, or provisions, need be incorporated in the solution procedure to insure that the solution satisfies the conservation of mass principle. Special procedures must be followed as described below, however, if PRV's are present in the network and the analysis is to be based on solving a system of equations in which corrective flow rates (the ΔQ's) are the only unknowns.

For networks without PRV's each equation in the system of ΔQ equations is obtained by summing the head losses around the loop through which that particular corrective flow rate ΔQ is thought of as circulating around. That is the mth equation will contain the mth ΔQ in each term of the equation. This procedure always automatically produces exactly as many independent equations as there are unknown ΔQ's for which a solution is sought. When PRV's are present this procedure of using identically the same loops for the corrective flow rates and energy equations must be altered. Reasons why alterations are necessary are: First, the head drop across a PRV cannot be expressed as a function of the ΔQ's circulating through that pipe. In fact, the head drop across a PRV is independent of the flow rate. Second, continuity at some junctions will not be satisfied if the ΔQ's are assumed to circulate through pseudo loops starting at artificial reservoirs created by the PRV's to another reservoir of the network. Reasons why continuity is not satisfied is that a ΔQ in such a pseudo loop would extract (add) fluid from (to) a junction, but not add (extract) an equal flow through another pipe joining at that junction.

Therefore, it is not practical to write energy equations around these loops which contain pipes with PRV's in them. Furthermore, it is not valid to let corrective flow rates, ΔQ, circulate through pseudo loops formed by a PRV at one of its ends without that pseudo loop first closing on the upstream junction of the pipe containing the PRV before proceeding to another resevoir. Forming such pseudo loops would add to the number of pipes in the loop and contribute to a larger banded Jacobian than necessary. Consequently, for networks containing PRV's some of the loops around which the energy equations are written cannot correspond to the loops around which the individual corrective flow rates, ΔQ, circulate. The individual ΔQ's will be assumed to circulate around the natural loops of the network, which are defined as if no PRV's existed. These ΔQ's will satisfy continuity at all junctions regardless of their individual values. The energy equations will be written around loops containing pipes or other elements such as pumps, or reservoirs whose head losses can be expressed in terms of the flow rate through them.

To illustrate this process, as well as to clarify how it must be implemented for different situations such as when the PRV exists in an external pipe, or in an internal pipe of the network, some very simple networks will be studied in detail. First, consider the simple network composed of seven pipes in the left sketch below. In this network the PRV is in an external pipe of the network. The natural loops consist of two in number and are composed of pipes 1, 2, 3, and 4; and 6, 7, and 2, respectively. These loops are determined the same as if the network contained no PRV's. The corrective flow rates ΔQ_1 and ΔQ_2 will

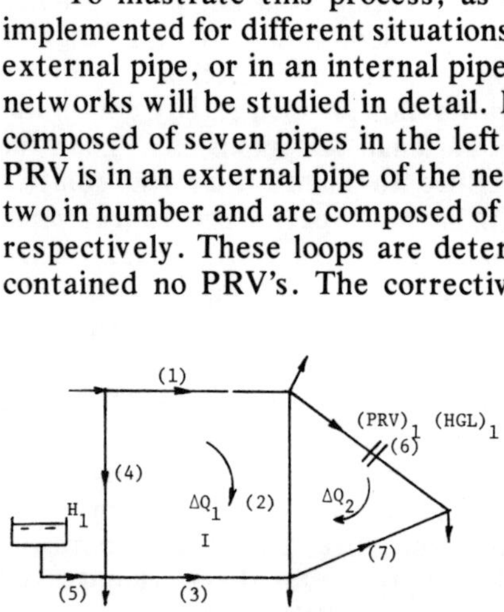

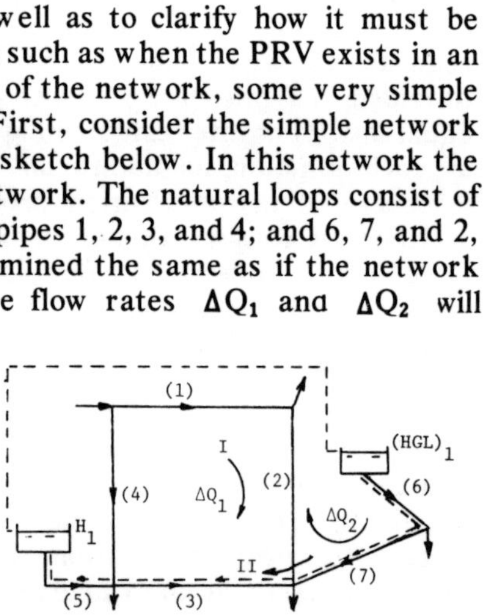

circulate through the pipes in these loops, respectively, as shown on the diagram. The energy equation can be written around loop I, but cannot be written around the second loop composed of pipes 6, 7, and 2, since the head drop across the PRV in pipe 6 cannot be expressed as a function of the flow rate through this pipe. The second energy equation will instead by written around the pseudo loop shown by the dashed line in the sketch on the right. This pseudo loop is found by disconnecting the pipe containing the PRV from its upstream junction and finding a series of pipes from the artificial reservoirs created by the PRV to the supply reservoir. This pseudo loop is composed of pipes 6, 7, 3, and 5. The two equations for solving for the two unknowns, ΔQ_1 and ΔQ_2 are:

$$F_1 = K_1 (Q_{01} + \Delta Q_1)^{n_1} + K_2 (Q_{02} + \Delta Q_1 - \Delta Q_2)^{n_2} - K_3 (Q_{03} - \Delta Q_1)^{n_3}$$
$$- K_4 (Q_{04} - \Delta Q_1)^{n_4} = 0$$

$$F_2 = K_6 (Q_{06} + \Delta Q_2)^{n_6} + K_7 (Q_{07} + \Delta Q_2)^{n_7} - K_3 (Q_{03} - \Delta Q_1)^{n_3}$$
$$- K_5 (Q_{05})^{n_5} + H_1 - (HGL)_1 = 0$$

in which the initial flow rates Q_{oi} satisfy the junction continuity equations. It should be noted in these equations that F_1 contains ΔQ_2 in the second term (which is associated with pipe 2) and that F_2 does not contain ΔQ_2 in the 3rd or 4th terms (which are associated with pipes 3 and 5, respectively).

Next, consider the same seven pipe network as above with the exception that the PRV exists in the internal pipe 2 instead of pipe 6. The corrective flow rates ΔQ_1 and ΔQ_2 circulate through the same natural loops as previously. Now, however, the energy equations are written around the loops shown by the dashed lines below. One of these is a real loop and the other a pseudo loop.

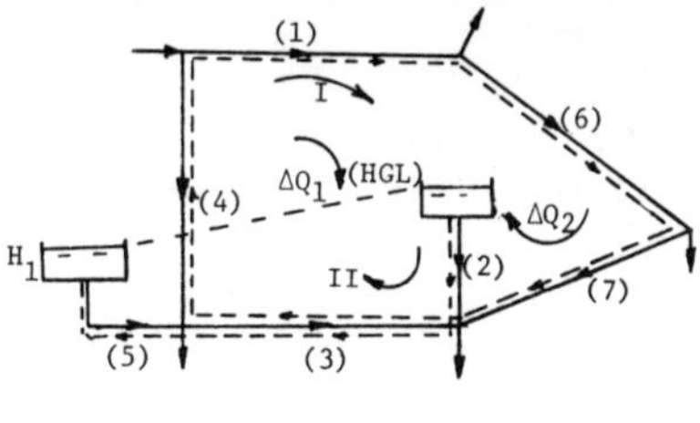

The equations which provide a solution to this network are:

$$F_1 = K_1 (Q_{01} + \Delta Q_1)^{n_1} + K_6 (Q_{06} + \Delta Q_2)^{n_6} + K_7 (Q_{07} + \Delta Q_2)^{n_7}$$
$$- K_3 (Q_{03} - \Delta Q_1)^{n_3} - K_4 (Q_{04} - \Delta Q_1)^{n_4} = 0$$

$$F_2 = K_2 (Q_{02} + \Delta Q_1 - \Delta Q_2)^{n_2} - K_3 (Q_{03} - \Delta Q_1)^{n_3} - K_5 Q_{05}^{n_5}$$

$$H_1 - (HGL)_1 = 0$$

It should be noted that the first of the above two equations is around the real loop, which encompasses the two loops around which the corrective flow rates circulate, and the second equation is for the pseudo loop containing the pipes 2, 3, and 5. This pseudo loop proceeds in the most direct route from the PRV to the supply reservoir.

The procedure to follow in setting up the system of ΔQ equations for networks containing PRV should now be unfolding in your mind. Should PRV's exist in two adjacent pipes one real loop for the energy equation encompasses the three natural loops around which the ΔQ's circulate, and two pseudo loops are formed starting at each of the PRV's. A portion of a network with two PRV's in adjacent lines is shown below. The extension of this principle to m PRV's in adjacent

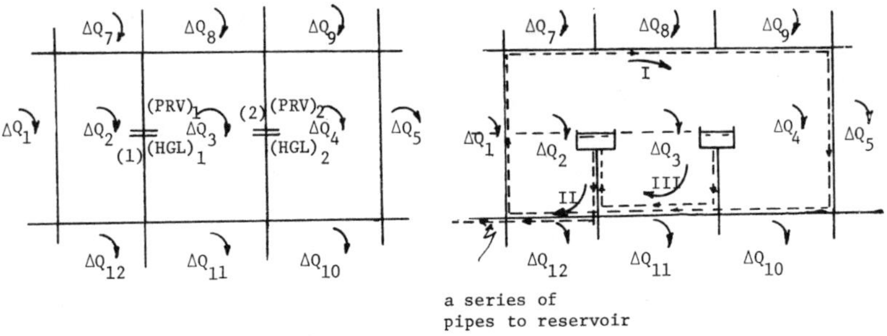

pipes is obvious. One energy equation is written around one real loop encompassing m + 1 natural loops around which m + 1 ΔQ's circulate. The additional m equations which are needed for a solution are energy equations from pseudo loops starting at the m PRV's.

PRV's acting as check valves

As described near the end of Chapter V, a pressure reducing valve will not permit the direction of flow through it to be opposite to its downstream direction. Rather under such conditions it acts as a check valve allowing the downstream pressure (or elevation of the HGL) to become larger than its pressure setting. When this occurs the ΔQ equations describing flow in the network must be altered as was necessary with the Q-equations used in the Linear Theory Method. The flow rate in the pipe containing the PRV is zero when a valve shuts off and, consequently, the corrective flow rate (or flow rates) for loops containing that pipe can be determined from the initial flow rates. If the pipe containing the PRV is external to the network and has only one ΔQ flowing through it, then that corrective flow is given by,

$$\Delta Q_j = \mp Q_{0i} \quad \text{.....................} (6\text{-}4)$$

in which the minus sign applies if the assumed directions for Q_{oi} and ΔQ_j are the same, and the plus sign if these directions are opposite to each other. Since ΔQ_j is now known, it is possible to replace it by the new unknown HGL downstream from the PRV such that the number of unknowns still equals the number of equations.

Should a PRV exist in an internal pipe which is part of two corrective flow rate loops, then one of the corrective flow rates is solved for in terms of the other one, and the HGL replaces this ΔQ in the list of unknowns. For example, in the previous seven pipe network with the PRV in pipe 2

$$Q_2 = (Q_{02} + \Delta Q_1 - \Delta Q_2) = 0$$

Therefore $\Delta Q_2 = Q_{02} + \Delta Q_1$

In every equation containing ΔQ_2 the substitution $Q_{02} + \Delta Q_1$ is made. The unknown (HGL), replaces ΔQ_2 in the list of unknowns. The two equations now needed to solve for the two unknowns ΔQ_1 and (HGL), are:

$$F_1 = K_1 (Q_{01} + \Delta Q_1)^{n_1} + K_6 (Q_{06} + Q_{02} + \Delta Q_1)^{n_6}$$
$$+ K_7 (Q_{07} + Q_{02} + \Delta Q_1)^{n_7} - K_3 (Q_{03} - \Delta Q_1)^{n_3}$$
$$- K_4 (Q_{04} - \Delta Q_1)^{n_4} = 0$$

$$F_2 = -(HGL)_1 - K_3 (Q_{03} - \Delta Q_1)^{n_3} - K_5 Q_{05}^{n_5} + H_1 = 0$$

This process of replacing ΔQ as they become known as PRV's shut off by the unknown HGL downstream from the PRV is similar for two or more parallel pipes containing PRV. For example, if the two PRV's in the two adjacent pipes shown below shut off, then

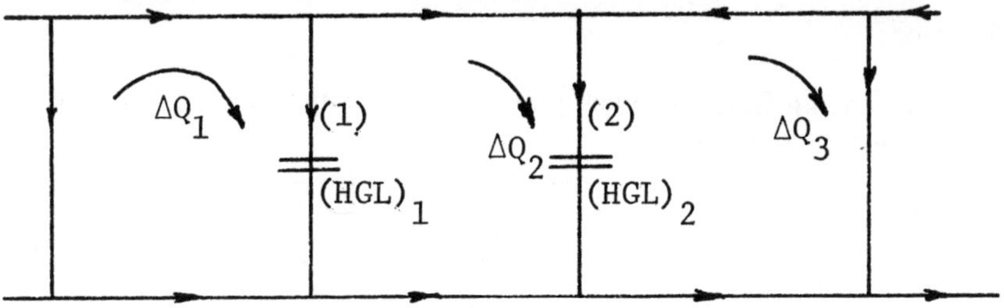

$$Q_1 = (Q_{01} + \Delta Q_1 - \Delta Q_2) = 0$$
$$Q_2 = (Q_{02} + \Delta Q_2 - \Delta Q_3) = 0$$

Since two HGL have now become unknown, they will replace two of the ΔQ's, say ΔQ_2, and ΔQ_3. Solving for ΔQ_2 and ΔQ_3 in terms of ΔQ_1 from the above equations give

$$\Delta Q_2 = Q_{01} + \Delta Q_1$$

and

$$\Delta Q_3 = Q_{02} + \Delta Q_2 = Q_{02} + Q_{01} + \Delta Q_1$$

Now it is possible to replace ΔQ_2 and ΔQ_3 wherever they occur in the ΔQ system of equations by known values (Q_{01} and Q_{02}) and ΔQ_1, allowing $(HGL)_1$ and $(HGL)_2$ to be considered unknowns in the system of equations. The extension of the known ΔQ's with unknown HGL's for three or more adjacent pipes containing PRV's which shut off follow the same process.

Example Problems in Solving ΔQ-Equations

1. Solve the 3-loop network below using the Hazen-Williams formula and the Newton-Raphson method to solve the ΔQ-equations.

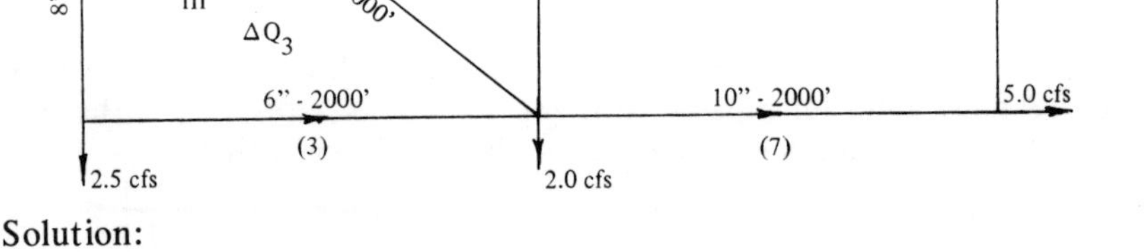

Solution:

Evaluating K_{HW} for each pipe as well as supplying an initial flow rate in each pipe gives,

Pipe No.	1	2	3	4	5	6	7	8
K_{HW}	2.074	3.78	60.65	11.21	14.94	30.32	5.04	7.56
Q_o (cfs)	5.0	2.0	2.5	5.0	3.0	0.5	4.5	2.0

The ΔQ-equations for this network are:

$$F_1 = 2.074\,(5+\Delta Q_1)^{1.85} + 3.78\,(2+\Delta Q_1-\Delta Q_2)^{1.85}$$
$$- 7.56\,(2-\Delta Q_1+\Delta Q_3)^{1.85} = 0$$

$$F_2 = 14.94\,(3+\Delta Q_2)^{1.85} + 30.32\,(.5+\Delta Q_2)^{1.85}$$
$$- 5.04\,(4.5-\Delta Q_2)^{1.85} - 3.78\,(2-\Delta Q_2+\Delta Q_1)^{1.85} = 0$$

$$F_3 = 7.56\,(2.0+\Delta Q_3-\Delta Q_1)^{1.85} - 60.65\,(2.5-\Delta Q_3)^{1.85}$$
$$- 11.21\,(5-\Delta Q_3)^{1.85} = 0$$

The solution to this system produces: $\Delta Q_1 = 0.712$, $\Delta Q_2 = -0.115$, $\Delta Q_3 = 2.24$, all in cfs. Therefore the flow rates and head losses in each pipe are:

Pipe No.	1	2	3	4	5	6	7	8
Q (cfs)	5.712	2.827	0.257	2.757	2.885	0.385	4.615	3.531
h_f (ft)	52.29	25.90	4.90	73.30	106.32	5.18	85.59	78.19

2. Solve for the flow rate in each pipe and the head at each junction of the network shown below. The pump characteristic curve is given by $h_p = -2.505\,Q_p^2 + 16.707\,Q_p + 155.29$ (h_p is in feet and Q_p in cfs)

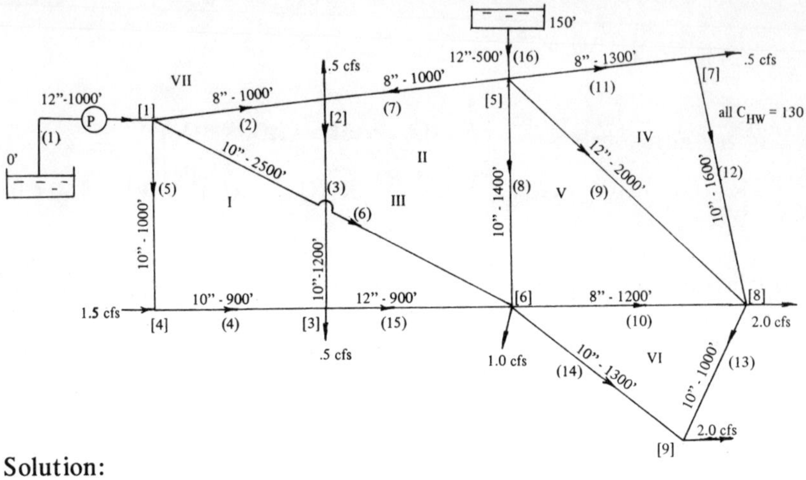

Solution:

Because of the pump and reservoir one pseudo loop connecting the reservoir by a no-flow pipe is needed. To solve for the seven corrective flow rates, seven energy equations are written as follows around the seven loops:

$$F_1 = K_2 (Q_{02} + \Delta Q_1 + \Delta Q_3)^{1.85} + K_3 (Q_{03} + \Delta Q_1 - \Delta Q_L)^{1.85}$$
$$- K_4 (Q_{04} - \Delta Q_1)^{1.85} + K_5 (Q_{05} - \Delta Q_1)^{1.85} = 0$$

$$F_2 = - K_7 (Q_{07} - \Delta Q_2 - \Delta Q_3)^{1.85} + K_8 (Q_{08} + \Delta Q_2 - \Delta Q_5)^{1.85}$$
$$- K_{15} (Q_{015} - \Delta Q_2)^{1.85} - K_3 (Q_{03} - \Delta Q_2 + \Delta Q_1)^{1.85} = 0$$

$$F_3 = K_2 (Q_{02} + \Delta Q_3 + \Delta Q_1)^{1.85} - K_7 (Q_{07} - \Delta Q_3 - \Delta Q_2)^{1.85}$$
$$+ K_8 (Q_{08} + \Delta Q_3 + \Delta Q_2)^{1.85} - K_6 (Q_{06} - \Delta Q_3)^{1.85} = 0$$

$$F_4 = K_{11} (Q_{011} + \Delta Q_4)^{1.85} + K_{12} (Q_{012} + \Delta Q_4)^{1.85}$$
$$- K_9 (Q_{09} - \Delta Q_4 + \Delta Q_5)^{1.85} = 0$$

$$F_5 = K_9 (Q_{09} + \Delta Q_5 - \Delta Q_4)^{1.85} - K_{10} (Q_{010} - \Delta Q_5 + \Delta Q_6)^{1.85}$$
$$- K_8 (Q_{08} - \Delta Q_5 + \Delta Q_2)^{1.85} = 0$$

$$F_6 = K_{10} (Q_{010} + \Delta Q_6 - \Delta Q_5)^{1.85} + K_{13} (Q_{013} + \Delta Q_6)^{1.85}$$
$$- K_{14} (Q_{014} - \Delta Q_6)^{1.85} = 0$$

$$F_7 = K_{16} (Q_{016} + \Delta Q_7)^{1.85} + K_7 (Q_{07} + \Delta Q_7 - \Delta Q_2 - \Delta Q_3)^{1.85}$$
$$- K_2 (Q_{02} - \Delta Q_7 + \Delta Q_2 + \Delta Q_3)^{1.85} - K_1(Q_{01} - \Delta Q_7)^{1.85} + h_p - 150 = 0$$

In the last of these equations h_p is defined by the pump characteristic curve equation with $Q_{01} - \Delta Q_7$ replacing Q_p. After supplying an initial flow for each pipe, and solving by the Newton-Raphson method the following solution results:

Pipe No.	1	2	3	4	5	6	7
Q (cfs)	4.975	1.653	0.114	2.777	1.277	2.045	-1.039
h_f (ft)	11.32	10.60	0.03	8.41	2.22	13.25	4.48

Pipe No.	8	9	10	11	12	13	14
Q (cfs) h_f (ft)	-0.960 1.83	1.415 2.21	0.889 4.04	0.609 2.17	0.109 0.04	0.413 0.27	1.587 4.31

Pipe No.	15	16
Q (cfs) h_f (ft)	2.391 2.62	0.025 0.00

Junction No.	1	2	3	4	5	6	7	8	9
Head (ft)	165.1	154.5	154.5	162.9	150.0	151.8	147.8	147.8	147.5

3. **Solve the network shown below. The wall roughness for all pipes is e = 0.012 inch.**

Solution:

Pipe No.	1	2	3	4	5	6	7
Q (cfs) h_f (ft)	19.80 10.59	10.37 2.97	4.89 0.69	4.00 24.57	2.66 0.22	4.13 17.43	4.44 10.04

Pipe No.	8	9	10	11	12	13	14
Q (cfs) h_f (ft)	4.60 16.17	13.67 11.30	2.40 3.01	4.07 16.90	6.07 2.29	1.64 2.86	1.11 1.01

Pipe No.	15	16	17	18	19	20	21
Q (cfs) h_f (ft)	5.47 0.86	1.60 4.10	0.40 0.10	1.33 1.45	2.83 0.52	4.83 1.02	4.07 12.69

Pipe No.	22	23	24	25	26	27	28
Q (cfs) h_f (ft)	2.59 10.46	1.41 1.07	3.07 7.29	3.11 15.00	3.07 7.28	2.57 5.16	6.98 61.39

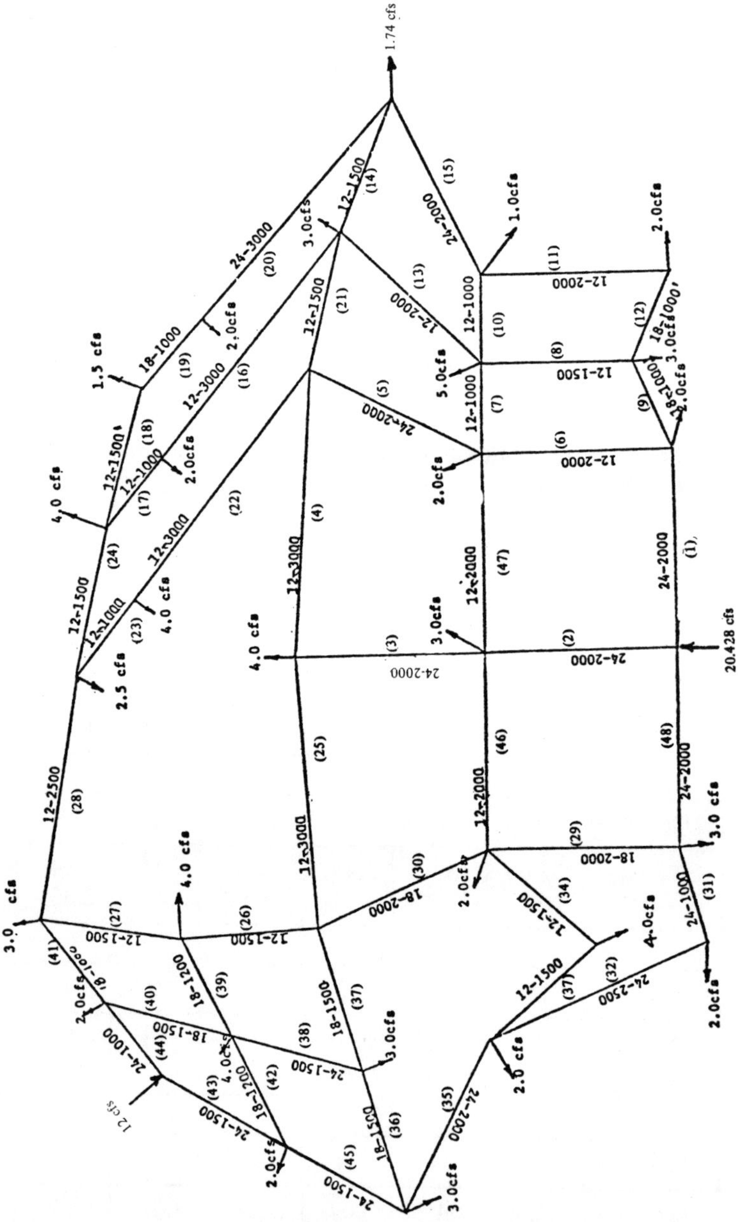

1.74 cfs
24-3000
(20)
12-1500
(14)
3.0cfs
24-2000
(15)
1.0cfs
12-2000
(11)
2.0cfs
12-2000
(13)
12-1000
(10)
12-1500
(21)
2.0cfs
18-1000
(19)
1.5 cfs
12-3000
(16)
5.0cfs
12-1500
(8)
(12)
3.0cfs
24-2000
(5)
12-1000
(7)
(9)
2.0cfs
12-1500
(18)
12-1000
(17)
2.0cfs
12-2000
(6)
4.0 cfs
(24)
12-3000
(22)
12-3000
(4)
12-2000
(47)
24-2000
(1)
12-1500
12-1000
(23)
4.0 cfs
4.0 cfs
(3)
3.0cfs
24-2000
(2)
20.428 cfs
2.5 cfs
12-2500
(28)
(25)
(46)
(48)
24-2000
12-3000
12-2000
18-2000
(29)
3.0 cfs
4.0 cfs
(30)
18-2000
2.0cfs
12-1500
(34)
24-1000
(31)
3.0 cfs
12-1500
(27)
12-1500
(26)
4.0cfs
(41)
18-1000
(40)
18-1200
(39)
18-1500
(37)
12-1500
(37)
(32)
24-2500
2.0cfs
2.0cfs
18-1500
(38)
24-1500
3.0cfs
24-1000
(44)
4.0cfs
(43)
18-1200
(42)
12 cfs
24-1500
2.0cfs
(45)
24-1500
18-1500
(36)
24-2000
(35)
2.0 cfs
3.0cfs

Pipe No.	29	30	31	32	33	34	35
Q (cfs)	2.48	8.02	10.26	12.26	2.94	1.06	17.21
h_f (ft)	0.81	7.89	1.46	5.16	6.73	0.92	8.03

Pipe No.	36	37	38	39	40	41	42
Q (cfs)	0.63	8.06	11.69	4.50	12.03	12.55	8.16
h_f (ft)	0.05	5.99	2.82	1.53	13.17	9.54	4.90

Pipe No.	43	44	45	46	47	48
Q (cfs)	29.73	26.58	19.58	2.48	4.96	9.74
h_f (ft)	17.75	9.48	7.77	6.41	25.04	2.63

4. **Solve the 48-pipe network of example problem 3 using the Hazen-Williams formula assuming the coefficient C_{HW} = 120 for all pipes.**

Solution:

Pipe No.	1	2	3	4	5	6	7
Q (cfs)	19.65	10.25	4.79	3.93	2.60	4.06	4.42
h_f (ft)	11.44	3.42	0.84	25.51	0.27	18.06	10.53

Pipe No.	8	9	10	11	12	13	14
Q (cfs)	4.58	13.59	2.39	4.01	6.01	1.61	1.09
h_f (ft)	16.87	11.72	3.37	17.64	2.59	3.23	1.78

Pipe No.	15	16	17	18	19	20	21
Q (cfs)	5.40	1.57	0.43	1.25	2.75	4.75	4.06
h_f (ft)	1.05	4.67	0.14	1.52	0.61	1.23	13.49

Pipe No.	22	23	24	25	26	27	28
Q (cfs)	2.48	1.52	3.18	3.14	3.04	2.47	7.20
h_f (ft)	10.88	1.46	8.60	16.83	7.93	5.39	65.07

Pipe No.	29	30	31	32	33	34	35
Q (cfs) h_f (ft)	2.41 0.95	7.94 8.66	10.07 1.66	12.07 5.79	2.97 7.57	1.03 1.07	17.04 8.78

Pipe No.	36	37	38	39	40	41	42
Q (cfs) h_f (ft)	0.41 0.03	8.04 6.65	11.44 3.15	4.57 1.87	11.93 13.81	12.67 10.29	8.09 5.38

Pipe No.	43	44	45	46	47	48
Q (cfs) h_f (ft)	29.72 18.45	26.60 10.02	19.63 8.56	2.50 7.33	4.96 26.07	9.47 2.96

5. Solve problem 4 with the demand at the junction of pipes 3, 4, and 25 increased from 4.0 cfs to 14 cfs; the demand at the junction of pipes 14, 15, and 20 decreased from 1.74 cfs to 0 cfs; the flow into the junction of pipes 43 and 44 increased from 56.312 cfs to 63.312 cfs; and the flow into the junction of pipes 1, 2, and 48 increased from 20.428 to 25.168 cfs.

Solution:

Pipe No.	1	2	3	4	5	6	7
Q (cfs) h_f (ft)	21.80 13.86	16.79 8.54	12.83 5.19	3.34 18.78	2.53 0.26	4.11 18.40	4.28 9.94

Pipe No.	8	9	10	11	12	13	14
Q (cfs) h_f (ft)	3.98 13.03	15.70 15.31	1.93 2.27	2.98 10.15	8.72 5.15	1.33 2.29	0.72 0.56

Pipe No.	15	16	17	18	19	20	21
Q (cfs) h_f (ft)	3.91 0.57	1.41 3.81	0.59 0.25	1.13 1.26	2.63 0.56	4.63 1.18	3.80 11.97

Pipe No.	22	23	24	25	26	27	28
Q (cfs)	2.07	1.93	3.46	4.50	3.89	3.00	7.89
h_f (ft)	7.75	2.28	10.06	32.77	12.48	7.70	77.20

Pipe No.	29	30	31	32	33	34	35
Q (cfs)	3.17	9.47	12.99	14.99	3.46	0.54	20.45
h_f (ft)	1.59	12.00	2.66	8.66	10.05	0.32	12.31

Pipe No.	36	37	38	39	40	41	42
Q (cfs)	0.76	10.08	13.84	4.89	13.68	13.89	9.05
h_f (ft)	0.08	10.12	4.48	2.12	17.79	12.21	6.63

Pipe No.	43	44	45	46	47	48
Q (cfs)	33.75	29.57	22.69	3.75	4.71	13.16
h_f (ft)	23.35	12.18	11.19	15.57	23.72	5.44

6. Using symbols Q_{oi} for initializing flow rates and K_i and n_i in the exponential formula to represent head losses, write out the system of ΔQ-equations which would be used to solve the networks in problem 11 at the end of Chapter V.

Solution:

$$F_1 = K_1 (Q_{01} + \Delta Q_1)^{n_1} + K_2 (Q_{02} + \Delta Q_1 - \Delta Q_3)^{n_2} - K_3 (Q_{03} - \Delta Q_1)^{n_3}$$

$$+ K_4 (Q_{04} + \Delta Q_1 - \Delta Q_2)^{n_4} - AG_1^2 - h_{01} = 0$$

$$F_2 = G_1 - (B/2A) = 0$$

$$F_3 = K_7 (Q_{07} + \Delta Q_2)^{n_7} - K_4 (Q_{04} - \Delta Q_2 + \Delta Q_1)^{n_4} - K_8 (Q_{08} - \Delta Q_2)^{n_8}$$

$$+ H_2 - H_1 = 0$$

$$F_4 = K_5 (Q_{05} + \Delta Q_3)^{n_5} - K_6 (Q_{06} - \Delta Q_3)^{n_6} - K_3 (Q_{03} - \Delta Q_1)^{n_3}$$

$$- K_8 (Q_{08} - \Delta Q_2)^{n_8} + H_2 - (HGL)_1 = 0$$

in which K_5 is determined for pipe 5 downstream from the PRV

7. Using symbols write-out the system of ΔQ-equations which would be used to solve the network in problem 13 at the end of Chapter V. (Note that PRV No. 2 in this network has shut off.)

Solution:

$$F_1 = K_2 (Q_{02} + \Delta Q_1 + \Delta Q_3 - \Delta Q_4)^{n_2} + K_3 (Q_{03} + \Delta Q_1 - \Delta Q_2)^{n_3}$$
$$- K_4 (Q_{04} - \Delta Q_1)^{n_4} - K_5 (Q_{05} - \Delta Q_1)^{n_5} - K_L (Q_{05} - \Delta Q_1)^2 = 0$$

$$F_2 = K_{10}(Q_{010} + \Delta Q_4)^{n_{10}} + K_0 (Q_{06} + \Delta Q_4 - \Delta Q_2 - \Delta Q_3)^{n_6}$$
$$- K_2 (Q_{02} - \Delta Q_2 + \Delta Q_1 + \Delta Q_3)^{n_2} - K_1 (Q_{01} - \Delta Q_4)^{n_1} + 50 = 0$$

$$F_3 = K_7 (Q_{07} + \Delta Q_2)^{n_2} - K_9 (Q_{09} - \Delta Q_3)^{n_9} - K_1 (Q_{01} - \Delta Q_4)^{n_1}$$
$$+ 300 - 150 = 0$$

$$F_4 = (HGL)_2 - K_7 (Q_{07} + \Delta Q_2)^{n_7} - 150 = 0$$

in which K_7 is determined for the length of pipe downstream from $(PRV)_1$.

8. Using the Newton-Raphson method to solve the system of ΔQ equations, obtain a solution to the 19 pipe network shown below.

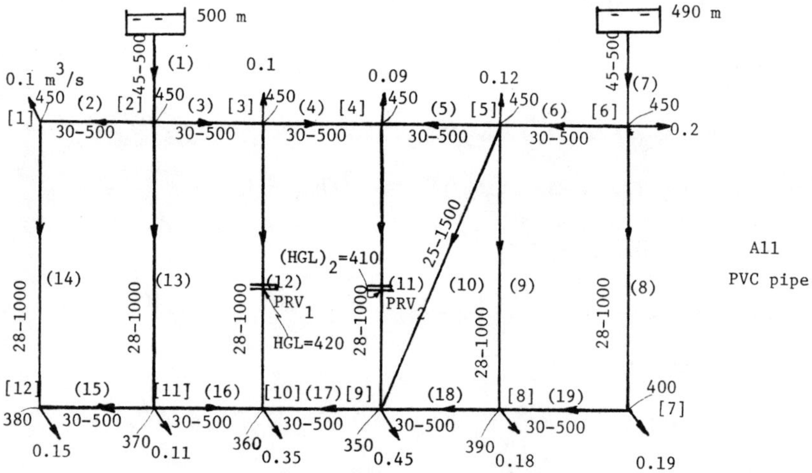

All diameters are in centimeters and all lengths in meters. Elevations are in meters and demands in m^3/s.

Solution:

Pipe No.	1	2	3	4	5	6
Flow rate (m^3/s) Head loss (m)	1.075 26.12	0.321 20.09	0.454 38.07	0.153 5.16	0.031 0.29	0.451 37.67

Pipe No.	7	8	9	10	11
Flow rate (m^3/s) Head loss (m)	0.965 21.38	0.314 53.89	0.182 19.73	0.118 23.32	0.094 5.94

Pipe No.	12	13	14	15	16
Flow rate (m^3/s) Head loss (m)	0.201 23.67	0.300 49.61	0.221 28.25	-0.071 1.28	0.261 13.73

Pipe No.	17	18	19
Flow rate (m^3/s) Head loss (m)	-0.112 2.91	0.126 3.60	0.124 3.51

PRV No.	1	2
Upstream, HGL (m) Downstream, HGL (m)	421.61 420.00	427.09 410.00

9. Obtain a solution to the network in problem 8 if the demand at junction 9 is decreased from $0.45\ m^3/s$ to $0.35\ m^3/s$.

Solution: $(PRV)_2$ has shut off.

Pipe No.	1	2	3	4	5	6
Flow rate (m^3/s) Head loss (m)	1.028 24.02	0.320 20.01	0.408 31.26	0.138 4.28	-0.048 0.63	0.406 30.91

Pipe No.	7	8	9	10	11
Flow rate (m^3/s) Head loss (m)	0.913 19.27	0.307 51.74	0.202 23.99	0.132 28.32	0.000 0.00

Pipe No.	12	13	14	15	16
Flow rate (m^3/s) Head loss (m)	0.170 17.39	0.299 49.35	0.220 28.09	-0.070 1.25	0.260 13.59

Pipe No.	17	18	19
Flow rate (m^3/s) Head loss (m)	-0.079 1.56	0.139 4.39	0.117 3.16

PRV No.	1	2
Upstream, HGL (m) Downstream, HGL (m)	434.29 420.00	440.45 411.49

Chapter VII
Hardy Cross Method

Introduction

The oldest and most widely used method for analyzing pipe networks is the Hardy Cross method, a description of which can be found in most hydraulics or fluid mechanics text books. In the precomputer days, hand solutions of pipe networks used the Hardy Cross method, and even today many computer programs are based on the Hardy Cross method. This method can be applied to solve the system of head equations, or the system of corrective loop flow equations, and for that matter no doubt also to solve the Q-equations. With the advent of the computer, as larger and more complex networks were analyzed, the Hardy Cross method was found to frequently converge too slowly if at all. A number of special measures have been employed to improve its convergence characteristics. These will not be discussed here. If a computer with sufficient storage is available, the methods described earlier are more suited. When using a small computer, or for hand computations of relatively small networks such as those given in the example problems in this book, the Hardy Cross method might still be used. In describing the Hardy Cross method in this book it will be used only to solve the ΔQ-equations. Its use in solving the H-equations or the Q-equations should be obvious from this discussion. The ΔQ-system is more frequently solved by the Hardy Cross method, probably primarily because it generally results in fewer equations. Most fluid mechanics books treat only its use in solving the ΔQ-equations even though this is generally not pointed out.

Mathematical Development

The Hardy Cross method is an adaptation of the Newton-Raphson method which solves one equation at a time before proceeding to the next equation during each iteration instead of solving all equations simultaneously. In doing this all other ΔQ's except the ΔQ_ℓ of the loop ℓ for which the equation is written are assumed temporarily known. Based on this assumption, the Newton-Raphson method can be used to solve the single equation $F_\ell = 0$ for ΔQ_ℓ, or

$$\Delta Q_\ell^{(m+1)} = \Delta Q_\ell^{(m)} - \frac{F_\ell^{(m)}}{\dfrac{dF_\ell^{(m)}}{d(\Delta Q_\ell)}} \qquad (7\text{-}1)$$

It is common in the Hardy Cross method to apply only one iterative correction to each equation before proceeding to the next equation (i.e. the next loop), even though several variations in detail exist. After applying one iterative correction to all equations the process is repeated until convergence is achieved. Furthermore, it is common to adjust the initially assumed flow rates in all pipes in the loop of that equation immediately upon computing each ΔQ. Consequently each equation $F_\ell =$ 0 is evaluated with all ΔQ's equal to zero, and furthermore the previous $\Delta Q_\ell^{(m)} = 0$. Consequently Eq. 7-1 reduces to,

$$\Delta Q = -\frac{F_\ell}{dF_\ell / d\Delta Q_\ell} \qquad (7\text{-}2)$$

The superscripts denoting iteration numbers in Eq. 7-1 are deleted in Eq. 7-2 because only one ΔQ appears. The equation $F_\ell = 0$ for this ℓ loop is the head loss equation around the loop or

$$F_\ell = \Sigma K_i Q_i^{n_i} \qquad (7\text{-}3)$$

The derivative of F_ℓ is,

$$\frac{dF_\ell}{d\Delta Q_\ell} = \Sigma \left| n_i K_i Q_i^{n_i - 1} \right| \qquad (7\text{-}4)$$

Substituting Eqs. 7-3 and 7-4 into Eq. 7-2 gives the following equation by which the Hardy Cross method computes a corrective flow rate ΔQ_ℓ for each loop of the network.

$$\Delta Q = -\frac{\Sigma K_i Q_i^{n_i}}{\Sigma \left| n_i K_i Q_i^{n_i - 1} \right|} \qquad (7\text{-}5)$$

If the Hazen-Williams equation is used to define K and n in the exponential formula, then Eq. 7-5 simplifies to

$$\Delta Q = -\frac{\Sigma (K_{HW})_i Q_i^{1.852}}{1.852 \, \Sigma \left| (K_{HW})_i Q_i^{0.852} \right|} = -\frac{\Sigma (h_f)_i}{1.852 \, \Sigma \left| (h_f/Q)_i \right|} \qquad (7\text{-}6)$$

The Hardy Cross method may be summarized by the following steps:

1. Make an initial guess of the flow rate in each pipe such that all junction continuity equations are satisfied.
2. Compute the sum of head losses around a loop of the network keeping track of signs. If the direction of movement (clockwise or counterclockwise) around the loop is opposite to the direction of flow in the pipe this h_f is negative. This step computes $\Sigma(h_f)_i$ or the numerator of Eq. 7-5 or Eq. 7-6.
3. Compute the denominator of Eq. 7-5 or Eq. 7-6 by accumulating the absolute values of $n_iK_iQ_i^{n_i-1}$ [or $(nh_f/Q)_i$] around the same loop.
4. Compute ΔQ by dividing the result from step 2 by the result from step 3, i.e. apply Eq. 7-5 (or Eq. 7-6).
5. Repeat steps 2 through 4 for each loop in the network.
6. Repeat steps 2 through 5 iteratively until all the ΔQ's computed during an iteration are small enough to be insignificant.

These steps are conveniently carried out in tabular form when doing a pipe network analysis by slide rule or desk calculator as illustrated below in solving for flow rates in the pipes of the two loop network given below. The solution will be based on the Darcy-Weisbach equation. The procedure will be simplified slightly if the Hazen-Williams equation is used. First, initial flow rates in each pipe are selected which satisfy the junction continuity equations. Based on these flow rates the next computations determine K and n in the exponeitial formula for each pipe as described near the end of Chapter II. The following table gives these values.

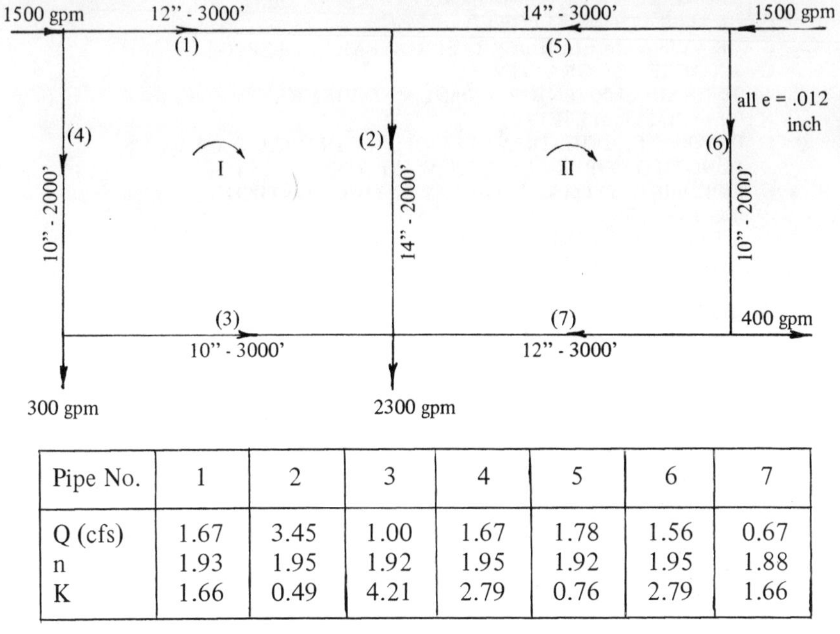

Pipe No.	1	2	3	4	5	6	7
Q (cfs)	1.67	3.45	1.00	1.67	1.78	1.56	0.67
n	1.93	1.95	1.92	1.95	1.92	1.95	1.88
K	1.66	0.49	4.21	2.79	0.76	2.79	1.66

From the values of K and n, the numerator and denominator in Eq. 7-5 (or Eq. 7-6) are computed as illustrated in the tables below.

Loop I

Pipe	Q	KQ^n	nKQ^n/Q
1	1.67	4.83	5.57
2	3.45	5.55	3.13
3	-1.0	-4.13	7.91
4	-1.67	-7.37	8.59
		-1.12	25.20

Loop II

Pipe	Q	KQ^n	nKQ^n/Q
5	-1.78	-2.29	2.46
6	1.56	6.45	8.04
7	0.67	0.76	2.15
2	-3.45	-5.55	3.13
		-0.63	15.78

$$\Delta Q_1 = -\frac{-1.12}{25.2} = 0.045 \text{ cfs} \qquad \Delta Q_2 = -\frac{-0.63}{15.78} = 0.040$$

Adding these ΔQ's to the flow rates gives: $Q_1 = 1.715$, $Q_2 = 3.455$, $Q_3 = 0.955$, $Q_4 = 1.625$, $Q_5 = 1.74$, $Q_6 = 1.60$, $Q_7 = 0.71$. After repeating the process several more iterations, the solution: $Q_1 = 1.767$, $Q_2 = 3.547$, $Q_3 = 0.984$, $Q_4 = 1.573$, $Q_5 = 1.862$, $Q_6 = 1.478$, $Q_7 = 0.588$, all in cfs, is obtained.

A listing of a simple FORTRAN program which uses the Darcy-Weisbach equation to determine head losses and obtains the solution by the Hardy Cross method is given below.

```
      REAL D(50),L(50),K(50),E(50),QI(50)
      INTEGER LP(42,7),NN(42)
C NP–NO. OF PIPES, NL–NO. OF LOOPS, MAX–MAX. NO.
C OF ITERATIONS ALLOWED.
C VIS–VISC. OF FLUID, ERR–ERROR PARAMETER, G2–
C 2 X ACCEL OF GRAVITY
   98 READ(5,100),END=99) NP,NL,MAX,VIS,ERR,G2
  100 FORMAT(3I5,3F10.5)
C II–PIPE NO., D(II)–DIAMETER OF PIPE IN INCHES, L(II)–
C LENGTH OF PIPE IN FT, E(II)–WALL ROUGHNESS OF
C PIPE, QI(II)–INITIAL FLOW SATISFYING CONTINUITY EQS.
      DO 1 I=1,NP
      READ(5,101) II,D(II),L(II),E(II),QI(II)
      E(II)=E(II)/D(II)
    1 D(II)=D(II)/12.
  101 FORMAT(I5,7F10.5)
      IF(G2 .LT. 1.E-5) G2=64.4
      ELOG=9.35*ALOG10(2.71828183)
      DO 51 I=1,NP
      QM=ABS(QI(I))
      DEQ=.1*QM
      AR=.78539392*D(I)**2
      ARL=L(I)/(G2*D(I)*AR**2)
      V1=(QM-DEQ)/AR
      V2=(QM+DEQ)/AR
      RE1=V1*D(I)/VIS
      RE2=V2*D(I)/VIS
      IF(RE2 .GT. 2.1E3) GO  TO 53
```

```
      F1=64./RE1
      F2=64./RE2
      E(I)=1.
      K(I)=64.4*VIS*ARL/D(I)
      GO TO 51
   53 MM=0
      F=1./(1.14-2.*ALOG10(E(I)))**2
      PAR=V1*SQRT(.125*F)*D(I)*E(I)/VIS
      IF(PAR .GT. 65.) GO TO 54
      RE=RE1
   57 MCT=0
   52 FS=SQRT(F)
      FZ=.5/(F*FS)
      ARG=E(I)+9.35/(RE*FS)
      FF=1./FS-1.14+2.*ALOG10(ARG)
      DF=FZ+ELOG*FZ/(ARG*RE)
      DIF=FF/DF
      F=F+DIF
      MCT=MCT+1
      IF(ABS(DIF) .GT. .00001 .AND. MCT .LT. 15) GO TO 52
      IF(MM .EQ. 1) GO TO 55
      MM=1
      RE=RE2
      F1=F
      GO TO 57
   55 F2=F
      BE=(ALOG(F1)-ALOG(F2))/(ALOG(QM+DEQ)-ALOG(QM-DEQ))
      AE=F1*(QM-DEQ)**BE
      E(I)=2.-BE
      K(I)=AE*ARL
      GO TO 51
   54 K(I)=F*ARL
      E(I)=2.
   51 CONTINUE
      WRITE(6,110)  (K(I),I=1,NP)
  110 FORMAT(' OCOEF. K IN EXPONENTIAL FORMULA',/,(1H  ,16F8.4))
      WRITE(6,111)  (E(I),I=1,NP)
  111 FORMAT('OEXPONENT N IN EXPONENTIAL FORMULA', /,(1H  ,16F8.4))
      DO 2 I=1,NL
C  NNP IS THE NUMBER OF PIPES AROUND THE LOOP
C  LP(I,J) ARE THE PIPE NO. AROUND THE LOOP. IF
C  COUNTERCLOCKWISE THIS NO. IS -
      READ(5,120) NNP,(LP(I,J),J=1,NNP)
  120 FORMAT(16I5)
    2 NN(I)=NNP
      NCT=0
   10 SUM=0.
      DO 3 I=1,NL
      NNP=NN(I)
      SUM1=0.
      SUM2=0.
      DO 4 J=1,NNP
      IJ=LP(I,J)
      IIJ=IABS(IJ)
      HL=FLOAT(IJ/IIJ)*K(IIJ)*QI(IIJ)**E(IIJ)
      SUM1=SUM1+HL
```

```
    4 SUM2=SUM2+E(IIJ)*ABS(HL)/QI(IIJ)
      DQ=SUM1/SUM2
      SUM=SUM+ABS(DQ)
      DO 3 J=1,NNP
      IJ=LP(I,J)
      IIJ=IABS(IJ)
    3 QI(IIJ)=QI(IIJ)-FLOAT(IJ/IIJ)*DQ
      NCT=NCT+1
      IF(NCT .LT. MAX .AND. SUM .GT. ERR) GO TO 10
      WRITE(6,105) (QI(I),I=1,NP)
  105 FORMAT(' FLOWRATES IN PIPES',/,(1H ,13F10.3))
      DO 9 I=1,NP
    9 D(I)=K(I)*ABS(QI(I))**E(I)
      WRITE(6,106)  (D(I),I=1,NP)
  106 FORMAT(' HEAD LOSSES IN PIPES',/,(1H ,13F10.3))
      GO TO 98
   99 STOP
      END
```

Appendix A
Units and Conversion Factors
Unit Definitions (SI)

1. *Meter*—m.

The meter is defined as 1650763.73 wave lengths in a vacuum of the orange-red line of the spectrum of krypton -86.

2. Time second—*sec*

The second is defined as the duration of 9192631770 cycles of radiation associated with a specified transition of the cesium atom.

3. *Mass kilogram*—kg

The standard for the unit of mass, the kilogram, is a cylinder of platinum-iridium alloy kept by the International Bureau of Weights and Measures at Paris.

4. Temperature *Kelvin*—°K

The Kelvin scale of temperature has its origin or zero point at absolute zero and has a fixed point at the triple point of pure water defined at 273.16°K. The triple point of the Celsius scale is defined at 0.01°C which is approximately 32.02°F on the Fahrenheit scale. Conversion between these temperature scales can be accomplished by the following equations

or

$$°F + 40 = 1.8\,(°C + 40)$$
$$°F = 1.8 \times °C + 32$$
$$°C = (°F - 32)/1.8$$
$$°K = °C + 273.15$$
$$°R = °F + 459.59$$

5. Force *Newton*—N

A force of 1 N when applied for 1 sec will give a 1 kg mass a speed of 1 m/sec. Gravity gives a mass a downward acceleration of about 9.8 m/sec^2 or 32.2 ft/sec^2.

6. Energy *Joule*

1 Joule = 1 N•m = 10^7 ergs = 0.239 cal
1 Btu = 778 ft-lb

7. Power

1 watt = 1 N•m/sec, 1 HP = 550 ft-lb/sec = 746 watts = 746 N-m/sec

BASIC CONVERSION UNITS		
	ES to SI	SI to ES
Length	1 in. = 2.54 cm 1 ft. = 0.3048 m	1 cm = 0.393701 in. 1 m = 3.28084 ft.
Mass	1 slug = 14.5939 kg	1 kg = 0.0685218 slugs
Force	1 lb = 4.44823 N 1 poundal = 0.138255 N	1 N = 0.224809 lb (1 N = 10^5 dynes)
SECONDARY CONVERSION UNITS		
Area	1 in.^{2} = 6.4516 cm^2 1 ft.^{2} = 0.092903 m^2 1 ac. = 0.4047 Hec (1 ac. = 43,560 ft.^{2})	1 cm^2 = 0.155 in.^{2} 1 m^2 = 10.76391 ft.^{2} 1 Hec = 2.471 ac. (1 Hec = 10,000 m^2)
Volume	1 in.^{3} = 16.3871 cm^3 1 ft.^{3} = 0.028317 m^3 1 gal. = 3.785 liters 1 fluid oz. = 29.57 cm^3 (1 ft.^{3} = 7.481 gals)	1 cm^3 = 0.061024 in.^{3} 1 m^3 = 35.31467 ft.^{3} 1 liter = 0.0353 ft.^{3} 1 liter = 1.057 quarts (1 liter = 1000 cm^3)
Density, ρ	1 slug/ft.^{3} = 515.363 kg/m^3	1 hg/m^3 = 0.0019404 slug/ft.^{3}
Specific wt.,γ	1 lb/ft.^{3} = 157.07 N/m^3	1 N/m^3 = 0.00637 lb/ft.^{3}
Pressure	1 psf = 47.877 Pa (N/m^2) 1 psi = 6894.24 Pa (N/m^2) (1 psi = 144 psf) (1 atm = 14.699 psi) (1 bar = 14.5 psi)	1 Pa = 0.0209 psf 1 Pa = 0.000145 psi (1 bar = 10^5 Pa) (1 Pa = 1N/m^2)
Dynamic Viscosity	1 lb-sec/ft.^{2} = 47.88 kg/m sec	1 N sec/m^2 = 0.0209 lb-sec/ft.^{2} (1 poise = 1 dyne sec/cm^2 = 0.1 N sec/m^2)
Kinematic Viscosity	1 ft./sec^2 = 0.0929 m^2/sec 1 ft./sec^2 = 929.02 stokes	1 m^2/sec = 10.764 ft./sec 1 stoke = 0.00108 ft.^{2}/sec (1 stoke = 1 cm^2/sec)
Energy, Work, Heat, Torque	1 ft-lb = 1.3558 Nm 1 ft-lb = 0.324 cal 1 Btu = 252 cal	1 Nm = 0.7376 ft-lb 1 cal = 0.00397 Btu 1 Btu = 778 ft-lb

Appendix B
Areas From Pipe Diameters

Diameter			Area			Diameter			Area		
in	ft	cm	in²	ft²	cm²	in	ft	cm	in²	ft²	cm²
¼	0.0208	0.635	0.0491	0.000341	0.317	5	0.417	0.7	0.63	0.1363	126.7
½	0.0417	1.27	0.1963	0.001364	1.267	6	0.500	15.24	28.27	0.1963	182.4
1	0.0833	2.54	0.785	0.00545	5.067	7	0.583	17.78	38.48	0.2673	248.3
1½	0.125	3.81	1.767	0.01227	11.40	8	0.667	20.32	50.27	0.3491	324.3
2	0.167	5.08	3.142	0.0218	20.27	9	0.750	22.86	63.62	0.4418	410.4
2½	0.208	6.35	4.909	0.0341	31.67	10	0.833	25.4	78.54	0.5454	506.7
3	0.250	7.62	7.069	0.0491	45.60	12	1.0	30.48	113.1	0.7854	729.7
3½	0.292	8.89	9.621	0.0668	62.07	18	1.5	45.72	254.5	1.767	1642.
4	0.333	10.16	12.57	0.0873	81.07	24	2.0	60.96	452.4	3.142	2919.
4½	0.375	11.43	15.90	0.1104	102.6	36	3.0	91.44	1018.	7.069	6567.

Diameter			Area			Diameter			Area		
cm	in	ft	cm²	in²	ft²	cm	in	ft	cm²	in²	ft²
1	0.3937	0.0328	0.7854	0.1217	0.000845	12	4.724	0.3937	113.1	17.53	0.1217
2	0.7874	0.0656	3.142	0.4869	0.003882	14	5.512	0.4593	153.9	23.86	0.1657
3	1.181	0.0984	7.069	1.096	0.007609	16	6.299	0.5249	201.1	31.16	0.2164
4	1.575	0.1312	12.57	1.948	0.01353	18	7.087	0.5906	254.5	39.44	0.2739
5	1.969	0.1640	19.63	3.043	0.02113	20	7.874	0.6562	314.2	48.69	0.3382
6	2.362	0.1969	28.27	4.383	0.03043	25	9.843	0.8202	490.9	76.09	0.5284
7	2.756	0.2297	38.48	5.965	0.04142	30	11.811	0.9843	706.9	109.6	0.7609
8	3.150	0.2625	50.27	7.791	0.05411	35	13.780	1.148	962.1	149.1	1.036
9	3.544	0.2953	63.62	9.861	0.06848	40	15.748	1.312	1257.	194.8	1.353
10	3.937	0.3281	78.54	12.17	0.08454	45	17.717	1.476	1590.	246.5	1.712

Appendix C
Description of Input Data Required by Programs Based on the Linear Theory Method in Solving the Q-Equations and the Newton-Raphson Method in Solving the ΔQ-Equations

Listing of the computer programs which require the input data described herein are not given, since these programs will likely be updated and improved with time, and a listing would occupy too much space. This description is given here since it is necessary to solve some of the example problems in the text. Also while the details might vary, the type of input data required by most computer programs will be similar to that below.

As background information to the input data the reader should be aware of the following:

1. Either English (ES) or international (SI) units may be used for data. If ES units are used pipe diameters may be given in inches or feet, pipe lengths in feet or 1,000 feet, and demands in gpm or cfs. If SI units are used, pipe diameters may be in centimeters or meters, and lengths in meters or 1,000 meters.
2. Frictional head losses may be based on the Darcy-Weisbach or Hazen-Williams formula.
3. All input data is free format, i.e., can be punched in any columns with a blank as the delimiter.
4. Besides pipes, the network can contain constant head reservoirs, source pumps, booster pumps, minor loss devices, and pressure reducing valves.
5. The program finds its own loops internally. No initialization is needed, but if desired an estimate of the fraction of inflow from supply reservoirs or source pumps may be supplied when using the program based on the Newton-Raphson Method.

The input data can be divided under the two general headings: (1) general specification, and (2) detailed data for each pipe. The following description is given under these two headings. Under "General Specifications" the card number will be used as a subheading. In many cases, particularly for large networks, the designated data may not all fit on a single card. Generally the data may or may not spill over on several cards. Therefore, a card number may refer to a sequence of cards.

General Specifications

Card No. 1

Column 2 contains the number of variations of the basic problem to be solved. If this is greater than 0, $CHANGE cards as described at the end of the data description follow the data for each variation specified. The remainder of this card contains any desired identification of this problem or problem sequence.

Card No. 2

This card begins with $SPECIF and ends with $END and contains any or all of the following variable names followed by an = and the required value. The separate variables are separated by commas. If the default option as indicated below is to be used, this variable does not need to be in the list. Only NPIPE is necessary.

NPIPE — No. of pipes in the network.

MAX — Maximum number of iteration allowed. (Default, MAX = 8 for linear theory program and 16 for Newton-Raphson program.)

NDIGIT — An accuracy indicator. The iterative solutions will be terminated when the sum of changes in flow rates (or corrective flows about loops when using the Newton-Raphson program) between consecutive iterations does not change in this digit to the right of the decimal point. (Default NDIGIT = 2.)

MINOR — Number of pipes containing devices which result in significant minor losses. (Default, MINOR = 0.)

NPRINH — If greater than zero, then values of the head at each junction of the network will be computed and printed. (Default NPRINH = 1.)

NPRINP — If greater than zero, then values of the pressure at each junction of the network will be computed and printed. (Default NPRINP = 1.)

IOEX — Level of intermediate data printed. -1 (least); 0 (intermediate); 1 or greater (most). (Default, IOEX = 0.)

IPROB — Problem specifications are printed out if equal to 1; these specifications are not printed if 0. (Default IPROB = 1.)

NUNIT — Type of units used for input data.
= 0 if dia. and wall roughness e in inches (or e equals the Hazen-Williams coefficient C_{HW}), and length in feet (or 1,000 feet).

= 1 if dia., wall roughness e, and length in feet (or 1,000 feet).
= 2 if dia., wall roughness e, and length in meters (or 1,000 meters).
= 3 if dia., and wall roughness in cm, and length in meters).

NPUMP — No. of pumps in the network. (Default NPUMP = 0.)

NBPUMP — No. of these pumps which are booster pumps, i.e., do not supply the network. (Default NBPUMP = 0.)

NRES — No. of reservoirs (or tanks) *without* pumps which supply the network. (Default NRES = 0.)

VIS — Kinematic viscosity of the fluid in ES units (ft^2/sec) or SI units (m^2/sec) according to problem units. (Default VIS = 0.00001217 ft^2/sec.)

GAMMA — Specific weight of the fluid in ES units (lb/ft^3) of SI units (N/m^3) according to NUNIT. (Default GAMMA = 62.4 lb/ft^3 if ES units used, = 9800 ft/m^3 if SI units used.)

NFLOW — Designates the units of the demands (or inflows) at the junctions. If units are gpm then NFLOW = 1; if units are cfs (or m^3/s if SI units are used, then NFLOW = 2.) (Default NFLOW = 2.)

NVALVE — No. of pressure reducing valves (PRV's in the network). (Default NVALVE = 0.)

NFREEF — If the free form of detailed data is *not* to be used, NFREEF should be greater than 0. If NFREEF = 3, then the fixed format of the detailed pipe data under the first form of this data input is read in on a punched card which precedes the pipe data cards. (See the discussion under the heading "Detailed data for each pipe".) If NFREEF is greater than 0 but not equal to 3, then the built-in fixed format is assumed. This format is: (I5, 3F10.5, I5, F10.5, I5, 2F10.5). (Default NFREEF = 0.)

COEFRO — Is the value of the roughness coefficient which will be assumed to apply for all pipe if no value is given on the pipe cards of detailed data input. (See the discussion under the heading "Detailed data for each pipe".) The value of COEFRO may be either the value of the wall roughness e if the Darcy-Weisbach equation is used or the Hazen-Williams coefficient C_{HW}. If only some pipes have the roughness coefficient equal to COEFRO, the remaining pipe may have their coefficient given on their respective pipe cards. If no values are given, the default value will be assumed and the Darcy-Weisbach formula will be used. (Default COEFRO = 0.0102 inches.)

NPLENG — If equal to 0, the length given for the pipes are assumed to be in feet if ES units are used, or meters if SI units are used. If NPLENG = 1, the lengths are given in 1,000 feet or 1,000 meters, depending upon the system of units. (Default NPLENG = 0.)

LOOPSE — If greater than zero then information must be provided, giving the pairs of pipe numbers at the end of each pseudo loop to be used. These data follows immediately the END which terminates the end of the pipe cards described below. The option of allowing the user to specify the ends of pseudo loops permits him to use considerable judgment which would be difficult to build into the program in forming loops which result in a minimum band width. Also, under rare circumstances, the order of selection of reservoirs, PRV's, or source pumps in forming pseudo loops which is used in the program may result in the selection of two such end nodes which are not joined together by any sequence of connected pipe. Should this occur, a message is written to the user (and the solution terminated) that he must examine the network carefully and assign LOOPSE = 1 and give the pairs of pipes at the ends of valid pseudo loops. Generally, however, this message is caused by errors in the input data, and the user should carefully examine the output printed to this point in the program to assist him in finding the error. (Default LOOPSE = 0.)

[NPERCT— This parameter is used only in the program based on the Newton-Raphson method of solution. If NPERCT = 1, then the fraction of the flow from each external supply source is required as described on Card No. 9a (default NPERCT = 1).]

(Additional variables NODES (No. of nodes), LOOPSE (No. of real loops), LOOPSS (No. of pseudo loops), etc. are also available that allow additional flexibility in use of the program, but are not discussed herein. For example, if NODES is given, an alternate form of input data is selected. Error terminations will result if LOOPSE and LOOPSS are given and the described network does not have the proper number of these loops.)

***Card No. 3** (Does not exist if default NRES = 0 is used.)

The pipe numbers which connect each reservoir (or tanks) to the network.

***Card No. 4** (Does not exist if default NRES = 0 is used.)

The elevations of the liquid surface in the reservoir (or tanks) in the same sequence as on the previous card.

***Card No. 5** (Does not exist if default NPUMP = 0 is used.)

The pipe numbers which contain the pumps. Numbers of pipes containing booster pumps must be listed *after* those which supply the network from wells or reservoirs.

***Card No. 6** (Does not exist if default NPUMP = 0 is used.)

For each pump in the network, a separate card is required defining the pump characteristic curve and the elevation of the liquid surface from which the supply is obtained. For booster pumps the elevation must be punched as 0.0. The pump characteristics are defined by giving flow rate (in cfs or m^3/s according to NUNIT) and the corresponding head produced by the pump for three points on the pump characteristic curve. The data on each card is Q_{p_1}, h_{p_1}, Q_{p_2}, h_{p_2}, Q_{p_3}, h_{p_3} and elevation. (This same information is required for booster pump with the elevation specified as 0.; the 0. must be punched.) Cards for all source pumps must precede cards for booster pumps.

***Card No. 7** (Does not exist if default NVALVE = 0 is used.)

The pipe number which contains the pressure reducing valves, PRV's.

***Card No. 8** (Does not exist if default NVALUE = 0 is used.)

The distances from the beginning of the pipe to the PRV's.

***Card No. 9** (Does not exist if default NVALUE = 0 is used.)

The elevations of the HGL immediately downstream from the PRV's.

[***Card No. 9a**

This card is required by the program which uses the Newton-Raphson method, but is not included in the program based on the linear theory program. In the same sequence as the reservoirs, PRV's and source pumps are numbered and their data is given above; this card contains an estimate of the fraction of the total inflow which comes from each such source. What is meant by "same sequence as reservoirs, PRV's and source pumps" is that the order of these fractions is first, the flow from reservoirs; second, the flow through lines containing pressure reducing valves; and third, the flow from source pumps. Within each of these three categories, the sequence is as given by the previous reservoir, PRV and pump cards.

If NPERCT = 0, this card is not required and the program initializes the flows in each pipe of the network under the assumption that the incoming flow is equally divided among all such sources.]

Card No. 10 (Fixed Formatted input)

(Only if the default of both no reservoirs, no pumps and no PRV is used does this card exist.)

This card establishes the elevation of the HGL at a designated junction. The data on this card is (1) the junction number, and (2) the elevation of the HGL at this designated junction. These data are right justified. Column 5 contains the designated junction number and in columns 6 through 15 the elevation of the HGL at this junction is given.

***Card No. 11** (Does not exist if MINOR = 0 is used.)

The pipe numbers which contain a device which causes a minor loss.

***Card No. 12** (Does not exist if MINOR = 0 is used.)

The coefficient K_L for the minor loss ($h_L = K_L (V^2/2g)$) in the sequence established on Card No. 11.

Detailed Data for Each Pipe

One and only one card is required for each pipe. Each such card contains the following data:

1. The pipe number NOP(I) (integer); 2. The pipe diameter D(I) (real); 3. The pipe length L(I) (real); 4. The roughness E(I) (real) (Hazen-Williams coefficient, or the wall roughness e of the pipe in inches, feet, meters or centimeters according to the specifications on Card No. 2). If omitted, the standard value given by the value assigned to COEFRO (or its default value e = 0.0102 inches) will be used; 5. The junction number from which the flow in this pipe comes, L11 (integer); 6. The external flow at the upstream junction QJ(L11) (real); 7. The junction number to which the flow in this pipe goes, L22 (integer); 8. The external flow at the downstream junction of this pipe, QJ(L22) (real); and 9. The elevation of the downstream junction, ELEV(L22) (real).

As is the case with the other input data cards, these data are free format, that is, the data do not need to be punched according to any format, i.e., in designated columns. However, if the data are punched according to some format, this format can be read in as input data before the first pipe card, if this has been specified by assigning the variable NFREEF = 3 on Card No. 2 as explained earlier. Also, the built-in fixed format may be used. Using a fixed format may save a small amount of computer time. However, the primary purpose for the feature is to make the program run on other computer systems which may not implement the special core read and write statements which are used for free format.

In using the free format option (i.e., NFREEF = 0), it is important that decimal points be punched in all real variables and not in integer variables. It is only by examining which values contain decimal points that the program allows some of the data for each to be omitted.

The external flow may follow the upstream or the downstream junction number. However, the elevation can only be given for the downstream junction number. Considering the variables on these pipe cards which may be omitted, any one of the following eight combinations of variables may be included on any pipe card. (In addition, E(I) may be deleted on any or all of these cards allowing 16 combinations.)

Case No.

1 NOP(I) D(I) L(I) E(I) L11 QJ(L11) L22 QJ(L22) ELEV(L22)
2 NOP(I) D(I) L(I) E(I) L11 L22 QJ(L22) ELEV(L22)
3 NOP(I) D(I) E(I) E(I) L11 QJ(L11) L22 ELEV(L22)
4 NOP(I) D(I) L(I) E(I) L11 L22 ELEV(L22)
5 NOP(I) D(I(L(I) E(I) L11 QJ(L11) L22
6 NOP(I) D(I) L(I) E(I) L22 ELEV(L22)
7 NOP(I) D(I) L(I) E(I) L11 L22
8 NOP(I) D(I) L(I) E(I) L22

These pipe cards do not need to be in the same order in which the pipes are numbered. Pipe numbers do not need to include all the integers from 1 to the total number of pipes. Neither is it necessary to number junctions consecutively.

To indicate the end of the pipe cards has been reached, a card should follow with the word END in columns 1, 2, and 3. This end card is necessary if any data follow the pipe cards.

Solutions to Variations of Problem

Upon completion of the solution to the network defined by the above data, solutions to variations of this basic network (i.e., same number of pipes, junctions, pumps, reservoirs, etc.) with pipes of different diameters, lengths or wall roughness or different junction demands, or junction elevations, as well as different reservoir elevations, can be obtained by giving such changes as described below. If one such variation of the basic network is to be used, the number in column 2 of Card No. 1 should be 1; if two such variations are to be solved this number equals 2, etc.

To specify each variation, a FORTRAN NAMELIST is used which contains the new pipe diameters D(I) in feet (or meters), the new pipe lengths L(I) in feet or meters and/or the new relative roughness e/D or E(I) under the NAMELIST name $CHANGE. For instance, to change pipe number 2's diameter to 0.5 feet, pipe number 10's length to 5,000 feet, pipe number 30's relative roughness to 0.015 and the demand at junction 5 to 0.7 cfs would require the following card.

$CHANGE (start in column 2) D(2) = 0.5, L(10) = 5000., E(30) = 0.015, QJ(5) = 0.7 $END

All changes must have units of feet or meters and cfs or m^3/s, except E(I), which is dimensionless.

Index

Areas, pipes 153
Artificial reservoir created
    by PRV 87
Bends 47
Bernoulli equation 6,11,14
Blasius 31
Branching pipes 56
Check valve 86,88,132
Colebrook-White equation 30,31
Computer programs 129
Conservation
    of energy 11
    of mass 6
    of momentum 23
Conservation laws 5
Consumption 60
Continuity equation 6,60,73
Conversion factors 151,152
Corrective flow rate equations
    67,124,134
Corrective flow rate loops 130
Corrective flows 67
Darcy-Weisbach equation 25,155
Data input 155
Density 1
Dimensions 2
Energy
    conservation 11
    kinetic 11,12
    potential 11,12
    total 11,12
Energy equations 11,14,63
Energy line 15
Energy loops 130
Energy per unit mass 11
Energy per unit weight 11
Enlargement 51
Equivalent sand roughness 31
Equivalent length 57
Equivalent pipe 54
Exponential formula 44
Flow rate equations or
    Q-equations 63,87,92
Fluid properties 1,3
Fluid shear 3
Fluid statics 16
Friction factor
    definition 25
    laminar flow 27,30
    turbulent flow 29,30
H-equations 65
Hardy Cross 54,71,145
Hazen-Williams coefficients 40
Hazen-Williams equation 40,155
Head
    elevation 14
    pressure 14
    total 15
    velocity 14
Head equations or H-equations
    65,117
Head loss equations 67

Horsepower 17
Hydraulic grade line (HGL) 15,89
Hydraulic radius 40
Impulse-momentum principle 22
Initialization 68,147
Input data 155
Junction continuity 60,62,73,87
Junction equations 65,87
Kinematic viscosity 3,4
Kinetic energy 11
Laminar flow 26
Linear theory method 71,87,155
Linear theory transformation 71
Loop equations 67,74,131,146
Loops 62,67,131
Loss coefficient 48,159
Loss of head
 due to bends 48
 due to contractions 50,51
 due to enlargements 51
 due to entrances 50,51
 due to exits 50,52
 due to fitting 47
 due to friction 25,40,41
Manning equation 41
Manning roughness coefficients 41
Mass flow rate 7
Minor losses 47,56
Momentum 22
Moody diagram 33
Newton force 151
Newton-Raphson method 37,54,115,125,155
Newtonian fluid 3
Node equations 65
Parallel pipes 55
Piezometric head 65
Pipe fitting losses 50
Pipes
 parallel 55
 series 54
Potential energy 11
Power 16,17
Pressure reduction valves 85,86,129,159
Pressure regulators 85,86,129
Pseudo loops 81
Pseudo loop equation 82,87
Pumps 17,80,82,159
Pump characteristics 82,159
Pump transformation 83,87
Q-Equations 67,87,92,155
Relative roughness of pipes 26
Reservoirs 80
Reynolds number 26
Roughness, pipe 31
Series pipes 54
Shear stress 3,27
Specific weight 2
Steady flow 6
Tanks 80
Transformation for pumps 83,87
Turbine 17
Turbulent flow 26
Units
 conversion factors 151
 Systeme International d'Unites (SI) 2,151
Velocity
 average 10,11
 gradient 3
Viscosity (Table 4)
 absolute 3,4
 kinematic 3,4
Volumetric flow rate 7,8
Wall roughness 31
Watt 17,152
Weight flow rate 7
Wetted perimeter 40
Wood, D. J. 40,71
Work 11
ΔQ-equations 67,145